FUZZY PLANNING

Introducing actor-consulting *as a means to address fuzziness in planning and decision-making*

And to harness the knowledge and mutual understanding of actors in an uncertain and complex world

Fuzzy Planning
The Role of Actors in a Fuzzy Governance Environment

Edited by

GERT DE ROO
University of Groningen, The Netherlands

GEOFF PORTER
Northumbria University, UK

Routledge
Taylor & Francis Group

LONDON AND NEW YORK

First published 2007 by Ashgate Publishing

2 Park Square, Milton Park, Abingdon, Oxon OX14 4RN
711 Third Avenue, New York, NY 10017, USA

Routledge is an imprint of the Taylor & Francis Group, an informa business

First issued in paperback 2017

British Library Cataloguing in Publication Data
Fuzzy planning : the role of actors in a fuzzy governance
    environment
    1. City plannng 2. Fuzzy decision making 3. City planning -
    Case studies
    I. Roo, Gert de II. Porter, Geoff
    307.1'216

Library of Congress Cataloging-in-Publication Data
Fuzzy planning : the role of actors in a fuzzy governance environment / edited by Gert
    de Roo and Geoff Porter.
        p. cm.
    Includes index.
    ISBN-13: 978-0-7546-4962-5
    ISBN-10: 0-7546-4962-8
    1. City planning--Research. 2. City planning--Case studies. 3. Regional planning--
Research. 4. Regional planning--Case studies. 5. Sustainable development. 6. Fuzzy
logic. 7. Uncertainty. I. Roo, Gert de. II. Porter, Geoff.

    HT165.5.F88 2006
    307.1'216--dc22

                                                                        2006021122

ISBN 978-0-7546-4962-5 (hbk)
ISBN 978-1-138-27534-8 (pbk)

Contents

Part C   Case Studies

List of Figures

List of Tables

Preface

This preface has been necessitated by the sudden death of my co-author, co-editor, appreciated colleague and friend Geoff Porter. As such it can also be regarded as an 'In Memoriam'.

Geoff Porter died of a heart attack on 30 August 2006, aged 53, and totally unexpectedly, while on holiday with his family, and while 'Fuzzy Planning' was still in the preparatory phases for publication.

I first met Geoff about seven or eight years ago. We were involved with an Interreg EU project on 'sustainable planning at the regional level'. Geoff was project leader, but rather new to the field. Three parties were involved in the collaboration. There were the Danes from Viborg whose basic interest was in farms, pigs and pig manure with an eye to controlling groundwater pollution. Then there were the Dutch from the province of Drenthe, who wanted to improve their already quite advanced, comprehensive regional plan. Finally there were the British from Newcastle upon Tyne, just there to learn. Geoff told me that the UK did not have a functioning regional body, one of the legacies of the Margaret Thatcher era. One thing I never quite got straight was whether it was the British who had a desire to learn, or just Geoff... The project ended two years later, with the Danes from Viborg still interested in only one thing – groundwater-polluting pigs – while the Dutch and the British felt there was actually more to it than that. I am not sure whether this was based on content or the fact that Geoff and I really enjoyed working together. Geoff in particular turned out to be a wonderful mediator. He was always able to consider the positive side of things. While I persisted in considering controlling the Viborg pigs as not the real issue in the project, Geoff constantly showed me their value as a case study. Once the project was over, we decided that we wanted to keep on working together and turn the Interreg project into a book. It has taken us four years to complete it, and here it is, two months after Geoff's death. We were able to find outstanding planners to contribute to this book, but it goes without saying that it was Geoff who invited our Viborg colleagues to join in too...

The book is called 'Fuzzy Planning'. The message in this book is to become aware of our preoccupations when considering various aspects of and in planning. The most profound example is sustainability. We all think we know what it means, but the moment it becomes operational we realise that that is far from true. 'Fuzzy Planning' tells us we have to understand that we are not as sure of anything as we think we are. That applies not only to planning but also to our individual lives. Real life is fuzzy, full of the unexpected!

For a long time we had a rather interesting working title for the book: 'Beware of Beautiful Days'. For marketing purposes that would not do at all, of course, but it was still a good title. Geoff loved it. It is a quote from a Nicci French novel, *The Red Room*, which goes like this: 'Beware of beautiful days. Bad things happen on beautiful days. Beware of having a plan. Your gaze is focused on the plan and that's the moment when things start happening just outside your range of vision'. Beware of having a plan. Geoff was full of plans, for his family, for this book, for having an academic career, to enjoy life. Nevertheless, he died of a heart attack, at the end of August 2006 while on holiday. It was totally unexpected – to my knowledge he was the only man in Newcastle-Gateshead who cycled to work! Beware of beautiful days as bad things can happen... There is an irony here, as you can see, and Geoff would have been the first to joke about it.

Geoff proposed the following preface: 'As a result of our growing awareness of the limitations of rational planning theory, an extensive discourse has developed on the subject of communicative planning theory. Nevertheless, it can be argued that both approaches have their uses. Communicative planning approaches tend to be useful in complex situations where many interests and conflicts come into play, whereas technical-rational approaches to planning can be used to solve the more simple problems and issues. But what happens between these two extremes?

This book argues that many of the key notions associated with spatial planning, such as 'sustainability' and 'compact city', are essentially fuzzy in their nature. The book introduces a method of data collection and analysis that, it is proposed, will help to clarify such situations, leading to the identification of more realistic spatial policy that reflects the thoughts, aspirations and motives of stakeholders or actors. This method is referred to as actor consulting.

The subject matter is introduced via a discussion of some of the key components of both communicative and rational planning: this includes models of governance and participation, and issues of scale, subsidiarity and decision-making. At the core of the book lies a theoretical description and reflection on the method of actor consulting. This tool, it is proposed, can be used to clarify, unpack or elaborate situations where fuzziness appears to be present. Worked examples and cases are then presented, which describe the use of the actor consulting model in addressing planning issues that are located in a variety of national-regional-local contexts, focusing in particular on the fuzzy notion of 'sustainability'.

The work on the theoretical concepts of this book and some of the cases presented within it originated in a transnational collaboration on spatial planning, supported by the Interreg IIC North Sea Region Programme, which seeks to develop trans-regional co-operation in spatial planning in the European Union. This book has been inspired by the ongoing co-operation in particular between the University of Groningen in the Netherlands and the University of Northumbria in the UK. Many contributions towards this book have however been made by recognised experts from across the world, and our thanks are extended to them for their input.'

It has been an honour to have had the chance to work with Geoff Porter. He became a personal friend and a good excuse to visit Newcastle upon Tyne. I would like to add my sincere thanks to all who participated in the Interreg IIc project, which started off this particular book on fuzziness. A warm vote of thanks is due also to the Province of Drenthe, who financed the translation costs of some of the chapters in this book. Thanks, too, to Tamara Kaspers for designing the maps used throughout the book. Valerie Rose deserves much appreciation for being a critical but supportive publisher, continually expressing faith in the 'Fuzzy Planning' project. And last but not least, many thanks to all the contributing authors for their patience and support.

Gert de Roo
Groningen

November 2006

Beware of beautiful days.
Bad things happen on beautiful days.
It may be that when you get happy, you get careless.

Beware of having a plan.
Your gaze is focused on the plan
and that's the moment when things start happening
just outside your range of vision.

Nicci French (2002) *The Red Room*, Penguin Books, London, p. 1.

Chapter 1

The End has no Merit...

Geoff Porter and Gert de Roo[1]

1.1 Introduction

> I am only interested in beginnings for there is so much hope at the beginning. The end has
> no merit except to demonstrate how badly that hope was misplaced (Walters 1994: 3).

In this book we focus on the idea that the language of planning is full of *fuzzy*
notions, concepts, doctrines, goals and visions. These are sources of uncertainty,
which until now have been highly underestimated by planners. The very fact that
many of the notions, concepts and such that underpin planning theory and practice
are indistinct or fuzzy by nature can often lead to false hopes and disappointing
outcomes to planning initiatives. Fuzziness in planning is the main theme throughout
the book. At the heart of our exploration of this theme, we present a tool – an actor-
consulting model – that is designed to address the fuzzy nature of planning. The
model is supported by theoretical arguments, and arguments that are based on
experience in practice.

As long as planners have the intention to control or to affect our daily environment,
by shaping and reshaping it according to the needs and wishes of the people, they
will be busy with *decisions,* and how to turn these into desired *effects.* Through
the years they have been confronted with undesired, unexpected and negative side
effects, as a by-product of their quest for intended outcomes. Nevertheless they have
learned substantially from the past about how decision-making processes work in
relation to an increasing variety and complexity of planning issues.

As a result, our more enlightened planning practitioners have reduced the use of
technical or functional approaches to a restricted variety of issues, in particular those
that appear to be straightforward. These are the issues we could define as 'simple'
due to the clarity of the features involved. For quite a number of issues however a
technical or functional approach leads to nothing but deceiving appearances. Our
national planning systems, with their origins rooted firmly in the technical-rational
thinking of the past, often continue to promote approaches to planning that fail to

---

1   The late Geoff Porter was a Researcher at the Sustainable Cities Research Institute,
Northumbria University, Newcastle upon Tyne, UK. Gert de Roo is Professor in Planning
at the Department of Planning and Environment, Faculty of Spatial Planning, University of
Groningen, Groningen, The Netherlands.   .

recognise the opportunities to solve complex problems with new tools (De Roo 2003).

The planning community has – just recently – embraced communicative approaches, to allow inter-subjective interaction and consensus to develop. Consensus can be seen as a means to cope with uncertainty in planning processes, in particular when numerous *actors* with widespread and diverse interests are involved. Both trial and error, and logical thinking have helped planners to develop a better understanding of decision-making processes. However, despite all efforts, good intentions and hopes, planners still face defeat on those occasions when results fail to work out as expected, do not emerge at all, display an unexpected twist, or become visible far later than might have been anticipated. Are we hoping against all odds to fully understand planning and decision-making processes? Will the frustration Walters shows us in the opening line remain the way of planning for the foreseeable future?

With this book we want to challenge this defeatism. To do this we will focus on those aspects of planning where against all expectations theory seldom seems to predict what happens in practice. To grasp these aspects in planning, we refer to major notions, concepts, doctrines, goals and visions such as 'liveability', 'participation' and 'compact city'. 'Sustainability' is considered to be one of the most significant fuzzy notions of our time, and as such is used here as a key example not only in this chapter, but throughout the book.

These high-level notions and concepts are often thought to be a guaranteed route to success, but all too often they result in disappointment. They are seemingly understood by all, and appreciated by all and therefore – one would think – accepted by all, in such a way that planning when turned into action will lead us to victory. But in practice, this is seldom the case...

We begin with the assumption that important notions, concepts and doctrines in planning, such as 'sustainability' and the 'compact city' are not understood as well as we implicitly might like to think. In this book we consider them to be essentially 'fuzzy', 'fluid' or 'illusive' by character. 'Sustainability' is perhaps the most obvious example: while almost everyone accepts 'sustainability' as one of the more important goals of planning, the outcome of the ambition to achieve 'sustainability' is often almost non-existent. One important reason is that its practical application is severely underestimated. For example, actors might inadvertently act in conflicting ways in the mistaken belief that their actions are contributing to 'sustainable development'. In this kind of situation the outcome of our planning might be minimal, or even in contrast to what was expected.

In this book we explore mechanisms that assist in understanding the fuzzy character of these concepts in planning, and how the actor is behaving accordingly. In particular, we propose a decision-making approach to tackle this fuzziness associated with these concepts and to clarify and to improve their operational value. A substantial part of this fuzziness may be the result of ignoring the fact that there is not one but many different understandings and perceptions of concepts such as 'sustainability'.

We therefore focus in particular on *actor-related fuzziness in planning*, and present a model based on the consultation and analysis of actor motives, perceptions and contributions (which we will call *actor-consulting*), as a tool to address differences in understanding and perception in planning. By inviting all crucial actors involved in a planning or policy scenario to express their *desired contribution*, their *actual contribution* and their *potential contribution* to – for example – the compact city concept, to liveability or to decentralisation decision-makers have a means to develop their goals in line with the role, motivation, perception and behaviour of the various actors involved. In this sense actor-consulting is not necessarily the same as public-consulting. As we will see later, for most of the planning issues the crucial actors to be consulted are likely to be governmental bodies, or third parties strongly affiliated to governmental policies. For a variety of issues in planning we believe this type of actor-consulting might contribute to the effectiveness, the efficiency and to the desired outcome of planning. This proposed actor-consulting model will also be challenged in this book. A variety of cases, located in Denmark, the Netherlands and the UK are presented, along with the lessons learned.

1.2   Fuzzy notions, concepts and doctrines

Planning practitioners and decision-makers are constantly under pressure to produce results and outcomes, often under circumstances that involve a high degree of conflict. Under such circumstances, it is understandable that there is little time to question the foundation of some of our basic concepts and doctrines, or to question the consequences of goals and objectives. Most of us have never given it much thought, or simply ignored signals that might be in contrast with our belief system. Part of the planner's language leaves room for interpretation. This does not only count for those notions in planning that are seldom used. On the contrary, some of the most notable notions, concepts and doctrines in planning have no clear and indisputable understanding. Most of us are willing to accept a rather vague conception as long as it corresponds with some sort of understanding. Only a troubled few find out they are in the dark the moment they attempt to turn such a concept into action, and that there is little space for discussion, because others believe there *is* a mutual understanding. So where do we go from here?

Let us explore for a moment the structure of our belief systems or terms of reference. A concept could be said to have become a doctrine when its underlying assumptions are accepted to such an extent that they are considered as implicit and given (Alexander and Faludi 1990, Korthals Altes 1995). Doctrines are therefore assumptions that have become prejudices and automatisms, and have become part of the basic principles that are broadly supported by society. Drior speaks of 'megapolicy assumptions' (in Alexander and Faludi 1990: 7): 'policy belief systems are of an order higher than the discourse about operational policies, plans, programs and/or projects'. These belief systems are not usually articulated, but they do influence the decision-making and planning processes. The term 'doctrine' refers to implicit beliefs that are more or less fixed in terms of the planning process (De Roo 1999, 2003). In this sense it is reasonable to say that notions such as for example 'sustainability' and 'urban' can be considered to be a doctrine on the basis that they are widely understood in society.

The point we want to make here is: should we take the implicit meaning of concepts and doctrines in planning for granted? And turn them into goals and include them in our objectives, just like that? Our answer is no, we should not. If we do not reconsider over and over again these notions, concepts and doctrines in planning, we rely on a certainty that simply is not there. If we fail to translate a notion, concept or doctrine into practice, we do not only fool ourselves. For example, decision-makers often delegate the responsibility for implementing a policy to an individual, or to an organisation, with insufficient powers to ensure that the desired outcome can actually be achieved. As a result our planning and decision-making have become little more than rhetoric. We have been defeated by a lack of clarity in our basic concepts and doctrines, or to put this in another way, by the fuzziness of the planning system. It is no surprise to see planning in this kind of situation as utterly useless.

Urban versus rural

Some understanding of fuzziness of the notions, concepts, goals and objectives we use in planning should not be entirely new to us. The debate about 'urban versus rural' begins with questions such as 'What is urban?', 'What is a city?', 'Where does the city end?' and 'Is rural the same as not-urban?'. These are classic debates in both planning and geography. In the first instance it seems to be a debate about the obvious. On reflection, though, it becomes highly frustrating to have to come to the conclusion that a clear and indisputable definition cannot be given, unless we agree on one. However such an arranged definition is no longer a reflection of reality, but a rather arbitrary description of a concept that we have decided to call 'urban'. What we perceive as 'urban' proves to be multi-functional and therefore highly complex to grasp with one all-embracing definition (Ashworth and Voogd 1990; Buursink 1980; Echenique 1976; Hall 2002; for 'rural' see Pacione 1984. See also Section 2.5 of this book where Healey speaks of the *multiplex city*). 'Urban' proves to be dynamic in time as well. For example, if in the past there was an understanding that 'urban' stood for relatively more intense interpersonal communication and interaction, the rise of mobile telephone networks may have diminished the significance of this notion. Planners are no longer trying to find universal definitions for 'urban' and 'rural'. They are defining these notions – when needed – specifically in relation to the issue they want to address. Indeed, the discussion has now moved on to other concepts, which are considered to have more practical application. Examples are the 'compact city', the 'city-region', 'ecological zones', 'functional regions' and so on.

In our quest to understand the fuzziness in today's planning systems we will continue here by focussing on 'the compact city', 'sustainability' and the way these two concepts relate to each other in an often contradictory way (De Roo and Miller 2000a). Concepts such as the 'compact city' are being used to express a desired spatial construction, and when seeking quality, the 'sustainability' doctrine is referenced. The compact city and sustainability are both examples of concepts that seem to grasp intentions in planning relatively well, however as demonstrated below, reality fools us because of its complexity, the consequent lack of a precise description of the terms we are using, and a biased belief system.

The compact city

In 1973 Dantzig and Saaty had already suggested the 'compact city' as the way to reduce urban expansion and preserve open rural areas. During that decade, larger cities in Western Europe actively started to reverse de-urbanisation in favour of the development of compact cities (Borchert, Egbers and De Smidt 1983). The 'compact city' is a spatial concept that is seen as a solution to various problems in urban areas, their direct surroundings and rural areas. It is however a concept that is not limited to spatial benefits alone. Indeed, concentration policies in France, Germany, the Netherlands and Scandinavia were initially motivated and driven by economic and social considerations (see RPD 1985). The potential of the compact city as a basis for economic and social development has always been considered a profound problem (Zonneveld 1991). Both the reduction of economic functions in some inner cities, and the consequent social imbalance that had developed in these areas was considered undesirable. The former has negative impacts on the attractiveness and the social safety of the inner cities – especially during the evening and at night. These socio-economic criteria – coupled with the intensification of the use of space, diversity and multi-functionality – were seen as essential characteristics to retain the urban basis (Jacobs 1961).

The compact city was believed to contribute to limiting the sprawl of urban centres, to reducing traffic, to improving accessibility, and – as a result – to improving the quality of urban life. However, when looking at events occurring in the real world, it is evident that dominant social developments are countering the effects promised by 'compact city' supporters. One of these developments is a demographic shift among households. Despite a relatively stable population, the number of households in most of Western Europe is rapidly increasing due to a rise of single-member households. This decrease in average household size necessitates a substantial restructuring program for housing.

An example is the Netherlands (see De Roo 2000, 2003). With a relatively stable population of seventeen million, residing in seven million dwelling units, another one million units must be built before the year 2010 (VROM 1999). The Dutch government decided to implement a guideline, which states that one third of the houses to be built should be allocated within the existing urban fringe, one third just beyond the existing outer-edge and one third at other locations (VROM 1993). This guidance however appears to be in contrast to public demand. Most households have a strong preference for single detached homes, with a parking space in front and a garden at the back of the house (Kempen 1994). A thriving economy, presenting people with the opportunity to spend more of their income on upgrading their housing standards adds to this demand. These factors lead to pressures to build high-value homes on spacious plots. Despite the compact city policy, there are therefore significant socio-economic pressures to encourage continuing urban sprawl for at least another ten years in the Netherlands.

It was also anticipated that the compact city would lead to a reduction in vehicular traffic (Le Clercq and Hoogendoorn 1983). Again economic demands have been among the many factors responsible for the opposite effect. The increase in road traffic and road building during the last quarter of the twentieth century has been relentless. The causes of this are complex, but are centred upon the (arguably misguided) perception across western society that the private car brings freedom, and that road freight is less expensive than other forms of transport (Vigar 2002). These factors have tended to undermine urban planning strategies centred upon the notion of the compact city. Our expectations that the compact city might lead to a reduction in travel have therefore proved to be too ambitious.

The 'sustainable city'

The belief in the possibility of a reduction in vehicular traffic has supported the idea that the compact city concept can contribute to sustainable development. Worldwide compact urban development has been pursued as a contribution to 'sustainability'. The report from the World Commission on Environment and Development, Our Common Future (1987), introduced the discussions on this subject, and numerous national governments have subsequently adopted sustainable development as a major policy element.

In Europe, the European Commission is a strong supporter of sustainable urban development. In its Green Paper on the Urban Environment, the European Commission (CEC 1990) lists a number of urban problems that obstruct sustainable development. It emphasises the fact that "dependency on motorised vehicles in general and private transportation in particular has increased" (1990: 48). According to the Commission, the burden on the environment in inner cities is partly the result of inner city congestion. Congestion leads to high levels of air and sound pollution. The Commission chooses several policy approaches, including a preference for multi-functional space and inner-city living; and a preference for solving urban problems within the existing city borders, avoiding peripheral expansion (CEC 1990: 45). This effectively embraces the compact city idea (also see the opinions of Breheny 1992 and Hall, Hebbert and Lusser 1993).

The UN Conference on Environment and Development held in Rio de Janeiro in June 1992 was a major impetus for 'sustainable' urban development (United Nations 1992). The UN considers a change of behaviour with regard to traffic and transportation as an essential step towards 'sustainable' cities. Efficient and ecologically sound transport systems for public transport and non-motorised means of transportation are considered important. However, the UN believes that practical details of the transition towards such progress should be determined at local and regional levels. Local governments are therefore expected to translate this programme of action to fit their circumstances: via Local Agenda 21 initiatives. Once again, it is believed that the concept of the compact city should provide opportunities for this translation (also see Jenks, Burton and Williams 1996).

Elkin et al. (1991: 12) imagine a sustainable city to "be of a form and scale appropriate to walking, cycling and efficient public transport, and with a compactness that encourages social interaction". Jenks, Burton and Williams (1996) subsequently conclude, "recently, much attention has focused on the relationship between urban form and sustainability, the suggestion being that the shape and density of cities can have implications for their future. From this debate, strong arguments are emerging that the compact city is the most sustainable urban form".

Compactness versus sustainability

The belief in the virtues of the compact city idea is widespread. Thomas and Cousins (1996) have summarised the benefits acclaimed by many (see e.g. Engwicht 1992, McLaren 1992, Newman and Kenworthy 1989, Sherlock 1991): "Less car dependency, low emissions, reduced energy consumption, better public transport services, increased overall accessibility, the re-use of infrastructure and previously developed land, the rejuvenation of existing urban areas and urban vitality, a high quality of life, the preservation of green space and a milieu for enhanced business and trading activities" (1996: 56). However, Thomas and Cousins also argue that these "claims are at the very least romantic and dangerous, and do not reflect the hard reality of economic demands, environmental sustainability and social expectations" (1996: 56).

Welbank (1996) concludes that the pursuit of sustainable urban development on the basis of compactness is primarily based on beliefs rather than on rational grounds (1996). Breheny, too, has doubts about the objectivity with which the idea of the compact city is treated as a sustainable concept, evidenced for example by a statement by the European Commission that dismisses suburban development as a 'sprawl' (Breheny and Rookwood 1993: 155). Jenks, Burton and Williams point out the weak relationship between the shaping of the concept and the foundations and assessment of the 'compact city' through research (1996: 7). Numerous authors take issue with the European Commission's view that identifies the 'sustainable city' with the 'compact city'.

The notion that compact cities would also be sustainable cities may therefore prove to be rooted in a belief in simplicity that may be non-existent. Cities are complex systems that respond insignificantly to predictions about factors such as reduction of mobility (De Roo 2003). A critical discussion of the simplistic point of view promoted by European Commission policy on urban planning and sustainability is thus essential.

So what is 'sustainability'?

After having reflected upon the concept of the compact city, the 'sustainable city' and interaction between compactness and sustainability, we now reflect upon 'sustainability'. This fuzzy notion is used throughout the book as an example of how various parties in planning interpret sustainability differently. Despite a good deal of common ground, a clear and indisputable definition of 'sustainability' is lacking.

In the UN Brundtland report, 'Our common future', 'sustainability' was presented as a concept combining economic, social and environmental aspects of growth and development (World Commission on Environment and Development 1987). The Brundtland report has proven to be a major statement, having significantly influenced the world arena of politics. Politicians and policy makers perceive sustainable behaviour as exercising care and respect in decision-making processes to offer future generations equal opportunities to use natural resources and to avoid behaviour that might lead to a heritage of pollution and disturbance. But, there is more to the situation. Contemporary societal issues vary in scale to those of the past. For example our environmental difficulties are manifest increasingly at the regional and global levels, because the practice of transferring pollutants to remote locations has become commonplace. Unfortunately such issues are often only recognised over the longer term. With sustainability we must not only respect future generations (time dimension), but the broader geographical scales at which environmental issues may arise (spatial dimension) (De Roo and Miller 2000b). Local decisions and local initiatives must therefore be seen as *contributions* to the achievement of sustainability at larger, national and global scales – hence the exhortation to 'think globally, act locally' (United Nations 1992).

The concept of sustainable development may be politically accepted, but the translation of this concept into practice is still cumbersome. A major reason for this difficulty is that sustainability is not something that can be implemented as such. Sustainability is a way of looking at policy-making and is – most of the time – included in the policy making process as a secondary objective. A second reason is its wide and confusing range of interpretations and its political and ethical character (O'Riordan and Voisey 1998). The OECD counted at least a hundred varying definitions of sustainability. Sustainability is cursed with fuzziness.

Studying sustainability and its implementation from an academic perspective therefore requires a somewhat different approach. Defining sustainability beforehand narrows down the scope of analysis, by prejudging its meaning (Lafferty and Meadowcroft 2000: 17). Instead, it should be seen as a policy concept that derives meaning by what people – actors – expect from it, and should "cut across established sectoral domains" (Lafferty and Meadowcroft 2000: 20). Analysing 'sustainability' should therefore focus on the intentions and belief systems of those who want to act or react to sustainability, whatever its meaning.

In practice it becomes essential to define sustainability in relation to the terms and characteristics of a particular project or initiative. Even then the way forward is still not apparent, as the understanding of the issue at hand varies from individual to individual. It is often found that not all actors are particularly interested in or in favour of 'sustainable development', especially if they are required to adapt their lifestyle (Davidson 1996). In terms of implementation therefore, a creative process is needed. Sustainability precludes a universal understanding of how it can be interpreted, and consequently we must regard its elaboration as a *fuzzy* planning problem that should be related to the motives, values and preferences of influential actors who have a direct interest in the planning issue at hand.

1.3 Understanding fuzziness

What does this analysis of the 'compact city' concept and the 'sustainability' doctrine tell us? We can conclude that two of the most revered notions in contemporary urban planning are hardly understood, despite the extensive amounts of energy put into them. While both notions are well intended, they are meant to be understood in general terms. This means that each concept is open to interpretation depending on the individual's position. Both are belief systems rather than reflections of reality. As a result they are lacking clear tools for implementation. Both also imply controlling our environment on the basis of prejudice, while facts are being ignored in favour of unrealistic ideals. 'Urban sprawl' and 'traffic congestion', to be solved by implementing the 'compact city' concept, are among the most obvious of the seemingly insoluble difficulties we are facing today. Perhaps we are not yet capable of understanding the complexity and dynamics of the real world, because they are fuzzier than we would like them to be? Do we have difficulty in differentiating reality from rhetoric?

Taking a constructivist's line of reasoning, it is important to understand that reality as we see it is a construction that each of us composes for himself, in continuous interaction with the outside world (object-oriented interaction) and through interactions with constructions made by others (intersubjective related interaction) (Scott 1995). Unless there is one clear definition accepted by all, there will be multiple interpretations, which will add to the uncertainty of a situation. Seen from a planning perspective, this adds to the uncertainty of the planning issue, to uncertainty within the planning process, and to uncertainty with regard to people's behaviour and actions.

From a philosophical perspective, Kosko tells us that fuzziness means *multivalence*: "three or more options, perhaps an infinite spectrum of options" (1994: 19), instead of one specific, unified meaning to address a single entity. Fuzziness arises therefore when a notion or concept has a multiple understanding. This suggests that we may need to review and clarify our understanding of even the most basic terms in planning.

We can conclude from these observations regarding the nature of uncertainty and fuzziness that notions, concepts and doctrines such as 'the compact city' and 'sustainable development' represent a vision that simply does not exist. Notions such as these must therefore be scrutinised and supported by facts and arguments.

Developing ways in which planning systems can be adopted in accordance with the outcome of this debate will bring us another step closer to grasping the reality that surrounds us.

1.4   Decision-making by actor-consulting

The good news is that there is a growing awareness among planners about the complexity and dynamics of the world around us. Planners are beginning to recognise the difficulties inherent in controlling or affecting our environment and shaping it according to our wishes. Accepting the complexity of the outside world will undoubtedly result in new planning approaches. This book describes one example of such a new approach: actor-consulting.

But firstly we must draw an important conclusion from the arguments presented so far. We have argued that commonly used notions and concepts in planning are to some extent fuzzy by character. There are perhaps two ways of responding to this conclusion. One is to find better notions, concepts and such, which are clear and indisputable, and have a substantial operational value as well. The other is to seek mutual understanding among the people who have to act in line with these fuzzy notions.

Let us examine the first of these two options. To find notions and concepts which are clear and indisputable to replace the more fuzzy ones is indeed an attractive alternative. However, this approach still breathes the air of 'full control', which – in a way – has been challenged by the arguments so far. Having the ability to exercise full control for every situation is a myth, due to the uncertainty, dynamics and complexity of the various circumstances to be faced. Furthermore, it is clear that there would be difficulty involved in replacing a notion or concept accepted by and imbedded in society by another one. And if we look closely at these fuzzy notions and concepts such as 'compact city' and 'sustainability', we will see that they represent not so much a crystal clear and an easily distinguishable object, but an intention. Fuzzy notions and concepts are also difficult to distinguish from each other. Often the notion or concept is part of a spectrum (consider the urban-rural continuum: see Nelissen 1974), leaving doubts about where precisely one part of the spectrum ends and another starts. Therefore an approach based on the re-definition of concepts and notions appears doomed to failure, although one is of course free to try.

The main argument of this book is therefore that when dealing with fuzzy aspects in planning, the practice of seeking mutual understanding among actors is more likely to have a positive effect on the planning process and outcome. It means that for each individual issue the notions, concepts and goals used have to be explained and discussed among the actors involved in that issue. Discussions with the actors will also reveal how these notions and concepts might be made operational. This book introduces a model that addresses the problem of fuzzy aspects in planning. The basic premise underlying this model is that in specific situations the meaning of notions, concepts and goals is constructed from a variety of thoughts and opinions, which by intention share much in common but may nevertheless lead to different attitudes among the actors involved. To work towards a common perception an appropriate level of consultation is needed (sometimes referred to as a 'learning process'), which in turn entails a continuous struggle and exchange of ideas over the definition and meaning of planning notions, concepts and goals. We believe that for a variety of planning issues an actor-consulting model could support the evolvement of such a common perception. Indeed the case studies carried out in the Netherlands and the UK (see Part C of this book) support the view that actor consulting reduces apparently conflicting views and helps to focus on common knowledge and understanding.

The aim of the actor-consulting decision-making model is to address the subjective nature of fuzzy notions and concepts in planning, to create a common understanding among actors, and to unravel underlying mechanisms that determine the actors' behaviour. The model therefore provides information about the thoughts and actions of actors. This information equips decision-makers with better anticipation in an uncertain policy arena. The model helps decision-makers to formulate well-considered, realistic policy by reducing this uncertainty. This actor-oriented approach goes beyond the concepts of 'participation strategies', 'collaborative planning', and 'communicative action', by providing information on actors' motivation, perception and behaviour. It might consequently help to develop a less uncertain planning environment.

Actions of actors do not come about independently. Their actions are determined by their own beliefs, desires and capabilities, and the institutional setting in which they act. All of these factors are dynamic because of interactions between actors and the development of new institutional arrangements. In such a context actors will act in a certain way – their *present contribution*. Meanwhile they have certain ideas about the way they want to act – their *desired contribution*. Additional literature study, expert meetings and further analyses can generate even more information about the *potential contributions* of actors to the planning process, and about internal and external conditions under which such contributions are possible (Figure 1.1). The basic model consists therefore of actively gathering information about the present, desired and potential contribution to the planning process of actors involved at the beginning of the planning process, in support of a realistic set of goals or a desired outcome. The argument here is that this decision-making model can contribute to a better understanding of how to cope with dynamics, complexity and uncertainty in planning, in order to contribute to progress and development. In that sense 'actor-consulting' is seen as a means to tackle actor-related fuzziness in planning.

1.5   Structure of the book

This book presents a critical reflection on *uncertainty* in spatial planning, expressed in terms of the *fuzzy* character that we can attribute to many planning issues. One of this book's intentions is to make planners aware of this issue. It also invites us to

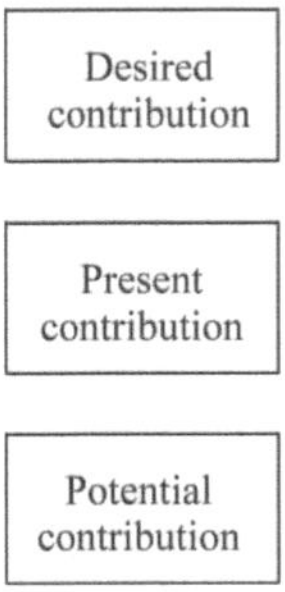

```
┌─────────────┐
│   Desired   │
│ contribution│
└─────────────┘

┌─────────────┐
│   Present   │
│ contribution│
└─────────────┘

┌─────────────┐
│  Potential  │
│ contribution│
└─────────────┘
```

Figure 1.1   The basis of the actor-consulting model; the desired, present and potential contributions (for more, see Figures 8.2 to 8.6)

search for a realistic approach to planning that takes into account this fuzziness and the complexity and subjectivity that comes with it. The book is structured in three parts. While we explore the limitations of notions, concepts, doctrines, goals and objectives in planning throughout the book, in Part A we particularly address the contextual setting of these fuzzy aspects in planning. We also have to understand that these contextual settings – the material world as it presents itself to us, and the institutional setting within which planning actions are framed – are not always well defined. The contextual setting is also a mental construction, which means that it will not be free from fuzziness either. Using sustainability as an example, the fuzzy nature of planning is explored from the perspective of these various settings. As such Part A gives us the context of our argumentation. Part B starts by positioning 'actor-consulting' within the current theoretical debate, from which it evolves as a decision-making model contributing to both planning theory and practice. In Part C, this model is tested using sustainability as an example in the context of a variety of policy issues in spatial planning in the UK, the Netherlands and Denmark. In doing so, we seek to explore how actor-consulting, coupled with changes to policy agendas might improve the outlook.

Part A   Contextual reflections

Part A presents a range of arguments from a variety of perspectives, supporting the view that certain aspects of planning are more fuzzy, illusive or fluid than we might expect. The nature of fuzziness is explored, supported by the suggestion that we should keep an open mind in complex planning situations. Four chapters focus respectively on different aspects of fuzziness in planning: fuzziness of notions and concepts; fuzziness of governance in relation to participation, communication and interaction of parties involved; fuzziness of intermediate or 'regional' levels in planning; and finally a reflection on these issues in relation to the use of indicators in planning.

In Chapter 2, Patsy Healey examines the fluidity of meaning of concepts in planning. Taking 'sustainability' as an example, she points out that in practice new concepts always undergo a transition in order to force a match with the prevailing institutional setting, and so as to fit within established policy arenas. However from a more positive perspective, she argues that the need to embrace a new concept demands a transformation of established institutions, discourses and policy relations. To sustain this transformative power, Healey argues that more attention is needed respectively to new ways of imagining objects of planning, to the way in which strategies are articulated, to the relations between strategy and action, and to the building of governance capacity.

Karel Martens is particularly interested in governance capacity and participation. His argument in Chapter 3 is that notions, concepts and doctrines not only are fuzzy, but that there is fuzziness encapsulated in the interactions of actors and the roles they play. He shows that the behaviour of actors is unpredictable, as is their motivation, and their perception of their environment. This is not only the case in policy arenas involving numerous actors. Most planning issues involve only a few parties, but even then the question of how they will actually behave in practice is often quite different from their stated position. This becomes even more relevant in the context that the traditional coordinative model of governance is being replaced by pluriform models of governance, resulting in a wide range of possible roles actors might be willing to play within the planning arena.

Henk Voogd and Johan Woltjer point out in Chapter 4 that intermediate levels in planning are clouded by fuzziness and fluidity. As an example the 'regional' level in planning is used, which they define as an intermediate level between state legislation and local planning and implementation. In the Netherlands, for example, these intermediate levels are meant to interpret guidelines and legislation from state level, but they also play a variety of unexpected roles, as well as being involved in conventional plans and objectives. Voogd and Woltjer show that even within a seemingly clear and well-defined policy system, with three layers of governance – state, province and municipality – the intermediate layer leaves room for interpretation, doubt and uncertainty.

In Chapter 5, Donald Miller emphasises the various roles indicators can play in planning, bearing in mind the context in which an indicator might be used. He illustrates the use of indicators within a variety of environments. In stable environments a technical rationality approach to planning is the obvious approach, which includes the use of content-bound indicators. However in unstable or *fuzzy* environments, either a pragmatic or a communicative rationality approach might be more useful, which also means a change in use of indicators. Here process-oriented and actor-oriented indicators, and the informative role these play, will be more helpful and will have better meaning than technical and content-bound indicators as a means to follow, to monitor, to examine and to evaluate the planning process and the progress it is making.

Part B   The actor-consulting model

Part B is centred on three chapters, the first of the three (Chapter 6) reflecting on
the arguments developed so far, and then reviewing these reflections in terms of the
history and theory of planning methods. A simple typology of planning is discussed
in Chapter 7, as a means to illustrate the opportunity for actor-consulting tools.
This typology is used to pinpoint where fuzziness in planning is likely to occur.
This leads to an explanation of fuzziness, and how it materialises on the interface
between 'complicated' and 'complex' planning issues. In Chapter 8, our method
for coping with fuzziness and fluidity in planning is then presented, focussing on
the role, behaviour, motivation and perception of the actors involved in a particular
policy arena.

The text emphasises that while content-oriented tools such as scenario and
contingency approaches are available, actor-oriented tools are in short supply. Our
actor-consulting model is particularly intended for use in situations in which the
number of actors participating is rather small, the formal role of the actors is more or
less known, and they act within a policy structure that appears to be well established.
In these circumstances both planners and actors can easily be led to believe that
the situation is fully understood, whereas in reality they are working in a fuzzy
policy arena and fuzzy environment. With actor-consulting, we hope to introduce
an approach that reduces this uncertainty, will add clarity to planning processes, and
last but not least will contribute to the outcomes of planning.

Part C   Case studies

Chapter 9 brings us to rural Denmark. Groundwater protection is of particular
importance in the County of Viborg, where nitrates easily penetrate the permeable
sandy soils. The main planning issue is the control of farming activities to prevent
long-term groundwater pollution, but difficult choices have to be made between the
socio-economic benefits of increased agricultural production, and their consequent
environmental effects. The environmental planners at Viborg, with strong political
support for the principles of 'sustainable development', have taken the initiative
to develop a geographic information system, as a means of stimulating public and
political debate. This allows citizens – not least the farmers themselves – to model
the possible effects of increased agricultural production, along with other relevant
parameters, on an internet-based system. Actor-consulting is used as a tool to analyse
the regulatory framework, and to justify the choice of GIS modelling as a means of
implementing sustainable development in regional planning policy.

Chapter 10 describes how actor-consulting was used to develop a policy
framework for the development and maintenance of cultural heritage in the historic
Wadden Sea Region, in the north of the Netherlands. The concept of 'policy life
cycles' was used by Rob de Boer in conjunction with this study, as a means to
analyse the contributions of actors over a protracted period of policy development,
implementation and monitoring.

In Chapter 11, Geoff Porter describes how actor-consulting was used as a vehicle to examine the sustainability of the local plan for the City of Newcastle upon Tyne, and of the effectiveness of regional planning policies. The case study examines the effectiveness of policy for housing location and design in the local and regional plans with regard to their ability to promote sustainable development.

Chapters 12 and 13 again use the Province of Drenthe as their study ground, where the revision of the Provincial Comprehensive Plan has sought to involve a wide range of actors in policy development, acknowledging that the implementation of the policy will require their mutual cooperation. In Chapter 12 Dana Kamphorst uses 'actor-consulting' as a tool to improve the planning goals for sustainable urban renewal. In the Netherlands urban renewal is primarily the responsibility of the municipalities, who depend on project developers and housing co-operatives to implement the local plans. The Province has the responsibility to support 'sustainable' urban renewal. As this can only be achieved in practice in a wider governance setting, this case study seeks to explore how the role of the Province can support the activities taking place. In Chapter 13 Rob de Boer uses 'actor-consulting' in the same institutional setting as Chapter 12, but within the policy sector of urban water management. Again the Province of Drenthe is seeking to improve cross-sectoral integration of policy by means of co-operation with the actors involved. The main actors are the Water Boards and the municipalities, both traditionally working independently from each other, but both now willing to explore the possibilities of working together, and both in doubt as to how to benefit from each others' expertise. The Province sees for itself a useful intermediate role, to the benefit of all, and to sustainability in particular.

The case studies described in both Chapters 11 and 14 focus on housing policy, and both seek to explore the effectiveness of the actor-consulting method. In Chapter 14, Geoff Porter and Frans Osté describe how a firm political commitment to 'sustainability' by the Province of Drenthe was translated into a range of policy recommendations for housing.

All these chapters together should give us answers to a number of questions. How can actor-consulting tackle fuzziness in planning? Will a form of planning emerge that is able to cope with increasing uncertainty and complexity? Perhaps the reader must come to his or her own conclusions, but whatever these may be, we hope that we have prompted some new thoughts and ideas as to how both academics and practitioners can play an active role in the quest for reaching a mutually acceptable outcome to planning activities.

References

Alexander, E.R. and A. Faludi (1990) 'Planning doctrine; Its uses and implications', paper for the conference on Planning Theory; prospects for the 1990s, Vakgroep Planologie en Demografie, Universiteit van Amsterdam, Amsterdam.

Ashworth, G.J. and H. Voogd (1990) *Selling the city; marketing approaches in the public sector*, Belhaven Press, London.

Borchert, J.G., G.J.J. Egbers and M. de Smidt (1983) 'Ruimtelijk beleid van Nederland; Sociaal-geografische beschouwingen over regionale ontwikkeling en ruimtelijke ordening' [Spatial policy of the Netherlands; Socio-geographical considerations about regional development and spatial planning], *De wereld in Stukken*, Unieboek B.V., Bussum (NL).

Breheny, M. (1992) 'Towards sustainable urban development', in: A.M. Mannion and S. Bowlby, *Environmental Issues in the 1990s*, John Wiley & Sons, Chichester, pp. 277–290.

Breheny, M. and R. Rookwood (1993) 'Planning the sustainable city region', in: A. Blowers (ed.) *Planning for a sustainable environment*, Earthscan, London, pp. 150–189.

Buursink, J. (1980) *Stad & Ruimte* [*City & Space*], Van Gorcum & Comp. BV, Assen.

CEC (1990) 'Green Paper on the Urban Environment', the European Commission.

Dantzig, G. and T. Saaty (1973) *Compact City; A plan for a liveable urban environment*, Freeman, San Francisco.

Davidson, M.D. (1996) *Grenzen aan duurzaamheid* [*Limits to sustainability*], Centrum voor energiebesparing en schone technologie, Delft (NL).

De Roo, G. (1999) *Planning per se, planning per saldo* [*Top down planning vs. bottom up planning*], SDU Publishers, The Hague (NL).

De Roo, G. (2000) 'Environmental conflicts in compact cities: complexity, decision-making, and policy approaches', *Environment and Planning B: Planning and Design*, Vol. 27(1), pp. 151–162.

De Roo, G. (2003) *Environmental Planning in the Netherlands: Too Good to be True; From Command and Control Planning to Shared Governance*, Ashgate, Aldershot (UK).

De Roo, G. and D. Miller (eds) (2000a) *Compact Cities and Sustainable Urban Development: A critical assessment of policies and plans from an international perspective*, Ashgate, Aldershot (UK).

De Roo, G. and D. Miller (2000b) 'Introduction – Compact cities and sustainable development', in: G. de Roo and D. Miller (eds) *Compact Cities and Sustainable Urban Development: A critical assessment of policies and plans from an international perspective*, Ashgate, Aldershot (UK), pp. 1–13.

Echenique, M. (1976) 'Function and form of the city region', in: T. Hancock (ed.) *Growth and change in the future city region*, Leonard Hill, London, pp. 174–189.

Elkin, T., D. McLaren and M. Hillman (1991) *Reviving the City; Towards Sustainable Urban Development*, Friends of the Earth, London.

Engwicht, D. (1992) *Towards an Eco-City, Calming the Traffic*, Envirobook, Sydney.

Hall, D., M. Hebbert and H. Lusser (1993) 'The Planning Background', in: A. Blowers (ed.) *Planning for a Sustainable Environment*, Earthscan, London, pp. 19–35.

Hall, P. (2002) *Cities of tomorrow; An intellectual history of urban planning and design in the twentieth century*, Blackwell Publishing, Oxford.

Jacobs, J. (1961) *The death and life of great American cities*, Vintage Books (edition 1992), New York.

Jenks, M., E. Burton and K. Williams (eds) (1996) *The Compact City, A Sustainable Urban Form*, Spon, London.

Kempen, B.G.A. (1994) *Wonen, wensen & mogelijkheden na 2000* [*Living, desires & options*] , Nationale Woningraad, Almere (NL).

Korthals Altes, W. (1995) *De Nederlandse planningdoctrine in het fin de siècle; Voorbereiding en doorwerking van de Vierde nota over de ruimtelijke ordening* (Extra) [*Netherland's planning doctrine in the 'fin de siècle'; Preparations and performance of the Fourth memorandum of spatial planning*], Van Gorcum, Assen.

Kosko, B. (1994) *Fuzzy Thinking: The New Science of Fuzzy Logic*, Flamingo, London.

Lafferty, W.M. and J. Meadowcroft (eds) (2001) *Implementing sustainable development: strategies and initiatives in high consumption societies*, Oxford University Press, Oxford.

Le Clercq, F. and J.J.D. Hoogendoorn (1983) 'Werken aan de kompakte stad' [Working on the compact city], in: *Planologische Diskussiebijdragen 1983, deel 1*, Delftse Uitgevers Maatschappij b.v., Delft (NL), pp. 155–166.

McLaren, D. (1992) 'Compact or dispersed?; Dilution is no solution', *Built Environment*, Vol 18(4), pp. 268–284.

Munda, G. (1995) *Multicriteria Evaluation in a Fuzzy Environment: Theory and Applications in Ecological Economics*, Physica Verlag, A Springer Verlag Company, Heidelberg, (D).

Nelissen, N.J.M. (1974) 'De stad; Een inleiding tot de urbane sociologie' [The city; An introduction in urban sociology], Sociologische monografieën, Van Loghum Slaterus, Deventer (NL).

Newman, P.W.G. and J.R. Kenworthy (1989) *Cities and Automobile Dependency, An International Sourcebook*, Gower Technical, Aldershot (UK).

O'Riordan, T. and H. Voisey (1998) *Agenda 21: the transition to sustainability*, Earthscan, London.

Pacione, M. (1984) *Rural Geography*, Harper & Row, London.

RPD (Netherlands National Planning Department) (1985) 'De compacte stad gewogen' [The compact city valued], Studierapporten RPD, nr. 27, VROM, The Hague (NL).

Scott, J. (1995) *Sociological Theory; Contemporary Debates*, Edward Elgar, Cheltenham, (UK).

Sherlock, H. (1991) *Cities are Good to Us*, Transport 2000, London.

Thomas, L. and W. Cousins (1996) 'The Compact City; A successful, desirable and achievable urban form?', in: M. Jenks, E. Burton and K. Williams, *The Compact City, A Sustainable Urban Form*, Spon, London, pp. 53–65.

United Nations (1992) 'Agenda 21; Programme of Action for Sustainable Development, Rio Declaration on Environment and Development, Statement of Forest Principles', The final text of agreements negotiated by Governments at the United Nations Conference on Environment and Development (UNCED), 3–14 June 1992, Rio de Janeiro, Brazil, United Nations Publications, New York.

Vigar, G. (2002) *The Politics of Mobility*, Spon, London.

VROM (Netherlands Ministry of Housing Spatial Planning and Environment) (1993) Fourth Memorandum on Spatial Planning Extra, Part 4 (TK 21.879 / nrs 65–66) Sdu Uitgevers, The Hague (NL).

VROM (Netherlands Ministry of Housing Spatial Planning and Environment) (1999) *Primos Prognose; De toekomstige ontwikkeling van bevolking, huishoudens en woningbehoefte* [Primos Forecast; The future development of population, households and need for housing], DGHV, VROM, The Hague.

Walters, M. (1994) 'The Scold's Bridle', Pan Books, Macmillan General Books, London.

Welbank, M. (1996) 'The Search for a Sustainable Urban Form', in: M. Jenks, E. Burton and K. Williams, *The Compact City, A Sustainable Urban Form*, Spon, London, pp. 74–82.

World Commission on Environment and Development (1987) *Our Common Future*, Oxford University Press, Oxford.

Zadeh, L.A. (1965) 'Fuzzy sets', Information and Control, Vol. 8, pp. 338–353.

Zonneveld, W.A.M. (1991) 'Conceptvorming in de Ruimtelijke Planning; Patronen en processen' [Conceptual developments in spatial planning; Patterns and processes], *Planologische Studies 9A*, Planologisch en Demografisch Instituut, Universiteit van Amsterdam, Amsterdam.

Part A
Contextual Reflections

Chapter 2

Re-thinking Key Dimensions of Strategic Spatial Planning: Sustainability and Complexity

Patsy Healey[1]

2.1 Introduction

In this chapter we examine the fluidity or fuzziness in the meaning of certain concepts in planning, by drawing on the perspective of institutionalist and interpretative policy analysis. In doing so, the concepts of 'sustainability' and 'sustainable development' will be discussed in particular, these being examples of notions or concepts mentioned in chapter one that have the potential to transform the policy agendas and policy processes involved in strategic spatial planning for cities and regions. We will illustrate that their translation into specific policy arenas, with their particular discourses and practices, could lead alternatively to transformation, or to a narrowing into the confines of established discourses and policy relations. To sustain transformative power, it is argued that more attention is needed to the key dimensions of 'sustainability' discourses. Three of these are studied: the nature of 'systemness', the nature of knowledge, and the nature of policy processes. The chapter concludes by arguing that, to transform the policy agendas and policy relations of strategic spatial planning, more attention is needed to new ways of reconsidering and actions of those involved in planning. Attention is also needed to new ways of imagining objects of planning such as cities and neighbourhoods, to the way strategies are articulated, to the relations between strategy and action and to building governance capacity.

2.2 A personal perspective

Those involved in discussions about 'sustainable development', 'sustainable cities', 'environmental sustainability' and 'sustainable environmental planning', come from a wide variety of backgrounds, encompassing different policy communities and different disciplinary traditions. As argued before in chapter one, and in this chapter

---

1 Patsy Healey is Professor Emeritus in the School of Architecture, Planning and Landscape, University of Newcastle, UK.

in more detail, there are few uncontested understandings and meanings to guide our encounters and we all have to learn how to listen to and grasp the meanings that others seek to convey in their arguments. It helps in this process for participants to explain the debates in which they have been involved.

This chapter is written from the perspective of a geographer-planner, trained to pay attention to the spatiality of social and natural phenomena, and specialising in the analysis of spatial planning systems and practices. This brings with it the perspective of a policy analyst, interested in the socio-political processes through which policy systems and interventions are designed and brought alive into social practices. The field of study has drawn on the development of debates in the field of planning theory, within which wider movements of thought have played out over the past twenty years. Within these movements, notions of policy design and development have been constructed and challenged through three major 'schools'. The *rationalist* school searches for objective principles through which policy goals could be translated into system design and specific interventions to achieve defined ends. The *structuralist* school examines the underlying conflicts between societal forces that define and constrain the role and design of specific policy systems and practices. The *communicative / interpretative* school explores the complex social processes through which meanings are created, lines of conflict and consensus are generated and policy systems invented, in terms of their formal expression and their social realisation. The author has been associated with the development of the third (Healey 1997), and has been especially interested in analysing the relations between the dynamics of policy process forms and the institutional contexts within which they evolve.

This chapter discusses the conceptual conflicts, which are highlighted by recent attempts to give meaning to the concepts of 'sustainability' and 'sustainable development' and the extent to which these have the power to transform the enterprise of strategic planning, and especially strategic spatial planning for regions, sub-regions and cities. The first section considers the nature of contemporary 'sustainability' policy discourses. The second section relates this general discussion to the adoption of 'sustainable development' as a primary goal in the British planning system and the struggles over the meaning this generates. Three areas of contest within these interpretative struggles are explored: the nature of 'systemness', the nature of knowledge and the nature of policy processes. The chapter concludes with some comments on the transformative potential of the concepts of 'sustainability' and 'sustainable development' within strategic spatial planning practices, and how these arguments contribute to the main theme of this book. Throughout, words in contemporary policy discourse whose meanings are contested are indicated by inverted commas.

2.3   The innovative force of 'sustainability' discourses

Whatever precise meaning we give to 'sustainability' as an idea, there can be no doubt of its influence in public policy in the past twenty years. Citing Brundtland,

Rio and some significant precursors, the concept is called up in an enormous range of academic, policy and pressure group literature. It has swept across environmental, urban and rural policy agendas and found a place in policy rhetoric in national and supra-national policy agendas across the EU and internationally.[2]

At a time when European policy design seemed to be stuck in sectoral and 'top-down' systems, the idea of 'sustainable development' called forward a new way of combining support for economic growth with much more attention to the long-term environmental and social consequences of such growth.[3] Closely linked to the mobilisation of concern for the natural environment associated with the rise of well-mobilised environmental lobby groups and 'green' political parties, it carried with it a revived policy interest in natural resource systems and ecological dynamics. This brought into policy debate an array of hypotheses about cause-effect relations, and hence of new policy foci and connections. It also revived older concepts of urban and environmental 'systems' and required explicit attention to ways of thinking about the scales of activity of 'systems', and of the significance of 'place' and 'locale'. It emphasised the importance of transforming the behaviour of people and organisations, and hence of the role of policy design and development in social learning and transformation processes. 'Sustainable development' as a policy concept thus seems to have the rhetorical power to act as a counterweight to the alternative late-twentieth century transformative agenda of market liberalisation. It seems to carry with it a re-valuing of environmental issues, and greater attention to social conditions and wider citizen involvement in policy development and design. Like the parallel concept of 'ecological modernisation' (Hajer 1995), sustainability discourses bring forward radical challenges. They focus on 'economic competitiveness' in a globalising economy, while moderating the radical edge in the hopeful assumption that economic, social, environmental and political concerns can be combined in some beneficial nexus. This can provide a strategic framing idea within which communities at many scales can move into a less-threatening future. The possibility of such 'holistic' integration is expressed in diagrams commonly found in planning documents, such as Figure 2.1.

Translating policy rhetoric into actual changes, in terms of the social dynamics through which urban and regional futures are produced, is never a simple linear process. Many policy analysts have questioned the extent to which new 'sustainability' rhetoric has become 'reality'. The very radicality of the ideas means that they challenge the established organisation of policy communities and academic disciplines. They challenge the traditional organisation of government in sectors, hierarchies and levels. They raise questions about the nature of the knowledge

---

2 Key references for UK and European debates are Haughton 1999; Innes 2000; Nadin et al. 2001; Owens 1997a; Owens 1997b; Ravetz 2000; Satterthwaite 1999; Selman 2000; Williams et al. 2000.

3 The literature contains various attempts to differentiate the concepts of 'sustainability', 'environmental sustainability' and 'sustainable development', but there is no consistency in usage or definition.

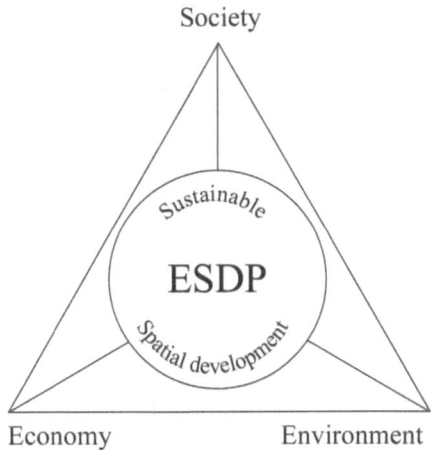

**Figure 2.1   An integrated policy agenda, as conceptualised in the European
                Spatial Development Perspective**
Source: Committee for Spatial Development 1999

resources which inform established government agendas and highlight the value
of the 'local knowledge' which citizens have about what is happening and 'what
works' in the places they are familiar with. This challenges the established power of
policy experts and professionals in translating policy ideas into practice. The policy
agendas raised by 'sustainability' discourses thus not only potentially transform
policy agendas. They generate a pressure to create new policy relations and practices.
As the rhetoric deploys notions of new 'holistic' approaches and 'more integrated'
management of local environments, so policy agendas and policy processes come
under pressure to co-evolve to create new policy 'frame'.

There can be no doubt of the innovative and transformative potential of
'sustainability' discourses. The notion has the rhetorical power to break up established
frames of reference and policy trajectories, bringing in new ideas and new actors.
But its power is also its weakness. As a rhetorical device, the term 'sustainability'
has a fluidity[4] of meaning.[5] As a broad banner behind which many new voices can
march into the citadels of government policy-making, this fluidity or fuzziness has
a positive benefit. But this power becomes a weakness when the storming of the
citadel has been successful, whereupon demands are made to translate the concept
into specific measures. The guardians of the citadel, having failed to keep out the
insurgents, may then pacify them by channelling their efforts into routines of policy
specification and control. Thus a huge amount of academic and policy development
effort is currently being deployed to develop techniques of impact assessment and

4   'Fuzzy' is the term used widely in this book to describe the fluidity of meaning of
many of our planning notions and concepts: see in particular Chapters 1 and 7.

5   And is not easy to translate into languages other than English, so the word itself is
often adopted into local languages

bundles of 'sustainability indicators' to assess progress in meeting 'sustainability' objectives (see Chapter 5). This ties down the specification of the meaning of 'sustainable development' into what can be 'objectively measured' across a wide range of contexts (Innes 2000, Sawicki 2002, Wong 2000, ECOTEC 2000). In the UK, it tames the innovative power of the discourse within the narrow confines of the 'audit culture', which serves to retain the power of central government over local initiative (Stewart 2002). This in turn may serve to alienate others from the rhetoric itself, particularly the pressure groups more remote from national policy culture and citizens more generally (Macnagthen and Urry 1998).

The fluidity of the concept of 'sustainability' and 'sustainable development' thus means that, in every institutional arena where attempts are made to give the concept operational meaning, struggles will take place over its interpretation. These struggles may be highly visible, but often they occur in the fine-grain of daily practices, as planners write texts of local plan policies, or municipalities define key indicators, or developers and resource managers frame their applications for regulatory approval. These struggles are about frames of reference (how to think about an issue), about who should be involved in policy design and implementation, and about the institutional arenas within which such efforts should take place.

2.4 Discursive struggle and English spatial planning

Maarten Hajer's pioneering work on 'discourse structuration' over the acid rain issue in Britain and the Netherlands in the 1970s and 1980s provides a vivid account of how the interplay between senior policy-makers and scientific factions was influenced by the policy cultures and histories of each national government (Hajer 1995). As a result, although there was a significant discursive shift in policy in both countries, in the Netherlands, the new perceptions led to quite fundamental changes in regulatory policy and social practices and widened the relevant policy communities involved. In the UK, the new ideas were contained within traditional policy communities.

But Hajer also argues that it is not enough to create a new policy frame and discourse. Discourses also have to be diffused, and used to shape practices (discourse institutionalisation). How is the sustainability discourse affecting the frames of reference and practices of the English planning system?

Sustainable development in the UK has been adopted as a banner for the planning system at national level for over a decade.[6] The power of the centre in the system is such that shifts in national policy agendas quickly flow into practice: into arguments by developers as to why they should be given planning permission, and into the texts of development plans (see, for example, Counsell 1999a, 1999b). Such plans (local plans, unitary development plans, structure plans and regional planning guidance) are now expected to have 'sustainability appraisals' to assess

---

6 'Sustainable development seeks to deliver the objective of achieving, now and in the future, economic development to secure higher living standards while protecting and enhancing the environment' (DoE 1997, paragraph 4).

their impacts (see for example the UK case study at Chapter 11), anticipating the
Strategic Environmental Assessment Directive (European Commission 2001). There
is no doubt that these innovations have had real effects on policies and practices.
The emphasis on developing 'brownfield sites' and 'safeguarding town centres'
has forced developers to re-assess their site acquisition strategies and the nature of
their development applications, if with varying degrees of willingness. The spatial
dispersal that was emerging from the 1980s as the planning regime was liberalised
has been reined in, justified by concepts of 'compact cities' and 'brownfield re-use'
and buttressed by the aims of 'reducing the need to travel', not wasting existing
infrastructure resources or consuming open land for development.

But these concepts of spatial organisation are hardly a radical departure for the
British planning system, which has been defending compact cities since the 1930s.[7]
They are linked to a strong and culturally embedded notion of the defence of the

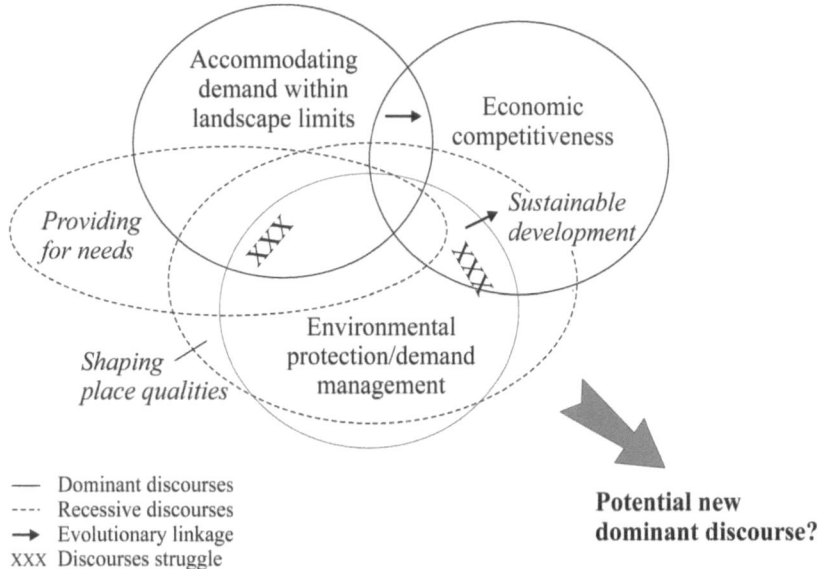

- — Dominant discourses
- ---- Recessive discourses
- → Evolutionary linkage
- xxx Discourses struggle

**Potential new
dominant discourse?**

Figure 2.2   Dominant and recessive discourses in British planning
Source: From Vigar et al. 2000, p. 243

countryside landscape, as embodying particular cultural values of 'Englishness'
(Healey and Shaw 1994). Thus the idea that cities, towns and villages should be
compact and not sprawled has an immediate political resonance, nationally and
locally. Developing different spatial concepts, such as 'transport corridors', receives
much less support. What the implementation of the 'sustainable development'

---

7   The old term was 'containment' (Hall et al. 1973).

discourse has done is to revive and re-configure these older notions, along with the dualistic notion of the role of the planning system as balancing 'economic' and 'environmental' considerations in the management of place qualities.[8] What so far has *not* been achieved is an expansion of the policy agenda to encompass social and cultural concerns about place quality within this dualistic agenda. Does this mean that the transformative power of the 'sustainable development' idea has been contained within established parameters, as with the acid rain controversy? Our research on strategic spatial planning in England in the 1990s strongly suggests that this is the case (see Figure 2.2).

Such containment is supported by the existence of some long-standing spatial planning concepts. One of these is the notion of the city or region as a 'system' (with a 'metabolism'), of the importance of 'comprehensive' strategies, and of policy cycles within which goals are translated into objectives and strategies, evaluated and monitored (for example, Ravetz 2000). Such ideas were introduced into planning theory and practice and into policy analysis generally in the 1960s and are still in wide currency across Europe. They have been challenged from the 1970s, for their techno-rationalism and for the reification of the expertise that they embody (see Chapter 6). Under the 'sustainable development' banner, there are many who continue to act as if experts and technical argumentation should dominate policy processes. But the context in which these concepts now operate is a very different one – intellectually, politically and in terms of urban and regional dynamics.

The radical edge of ideas marching under the 'sustainability' banner has been blunted in their encounter with established frames of reference and ways of implementing policy ideas in the British government culture. However, this culture is itself under pressure from wider shifts in knowledge and attitudes to governance, which are co-evolving with the concept of 'sustainable development'. In the area of spatial planning, this affects:

- the way locales, cities and regions are imagined;
- the processes through which strategies are created and articulated;
- the processes through which strategies invoke specific actions and have material and mental effects.

These shifts are slowly re-configuring policy agendas and concepts of how policy systems should be designed. As a result, concepts of 'sustainability' and 'sustainable development' are not merely struggling to retain their innovative 'edge' against established government cultures. They are also being re-shaped by pressures for transformation from other directions. This implies that, if 'sustainability' discourses are to have transformative force, attention needs to be given to the inherited baggage they carry within them. Policy change that transforms involves more than struggles over specific policy agendas (Cars et al. 2002).

---

8   This conception seems likely to underpin the coming debate on the reform of the planning system in England and Wales.

The next section explores some of this conceptual baggage and the challenges to it that are evident in the fields of planning and policy analysis, particularly with regard to cities and regions and their strategic spatial planning. The focus is on three issues: the nature of 'systemness'; the nature of knowledge; and the nature of policy-processes. The section concludes by assessing how these shifts are changing the way cities are imagined, how strategies are created, how they come to have material and mental effects and their overall relation to the transformative potential of the 'sustainability' agendas.

2.5 The nature of 'systemness'

The concept of 'system' is widely used in 'sustainability' discourses, where environments, cities, water resources, and ecological relations are regularly referred to as 'systems'. Justifications for more 'comprehensive', 'holistic' and 'integrated' approaches to environmental management are commonly grounded in the argument that the policy and practice relations of the 'management system' need to reflect the interrelated nature of the 'system' being managed. The great attraction of the 'system' concept is that it focuses attention on the ways 'parts' are connected to 'wholes', and hence the relational infrastructure of flows and feedback loops through which action in one time and place gets to have effects on another place and time (Figure 2.3). In the twentieth century, the idea flowed into many academic and policy arenas from a combination of inspirations from ecology and the science of missile control (von Bertalanffy 1970). In the 1960s, it swept across the planning field, providing a radical re-framing of concepts of the nature, purpose and method of planning.[9]

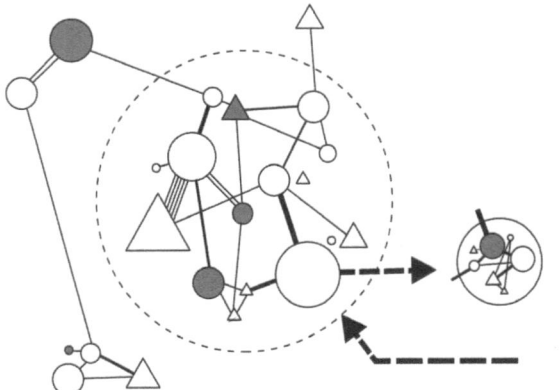

A system is a set if interconnected parts...

...but each part may be seen as a system itself...

... and the whole system may be regarded as but one of a larger system...

Figure 2.3   A conception of a system
Source: McLoughlin 1969, p. 76

---

9   Key contributors include: Chadwick 1970, Forrester 1969, McLoughlin 1969.

Instead of designing cities as physical artefacts, the role of planning was re-imagined as the guidance of change in urban systems. The planner was no longer 'building the city', but 'steering the city', encapsulated in McLoughlin's metaphor of the planner as the 'helmsman steering the ship'. In this conception, the urban system had an evolving trajectory, but sometimes got out of balance. The guidance job was to identify and correct these 'unbalancing' pathologies (Figure 2.4).

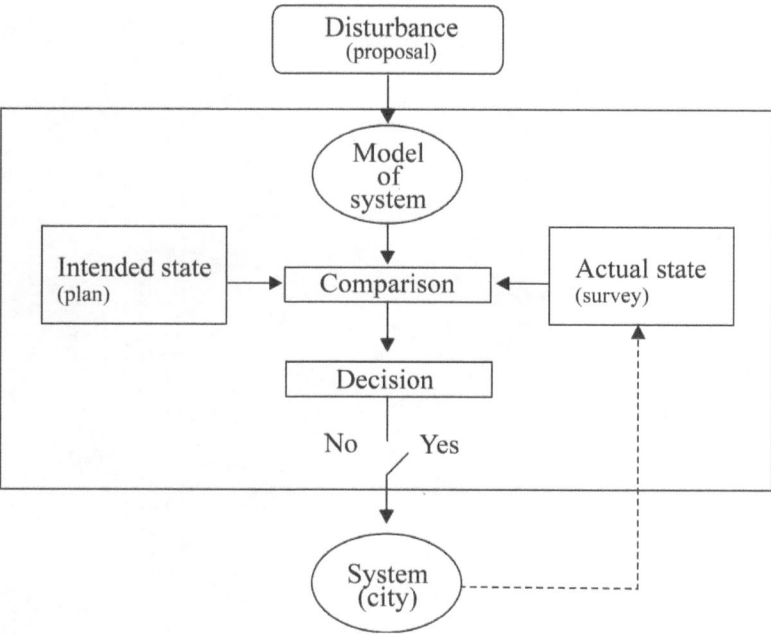

Figure 2.4   'Error-controlled regulation'
Source: McLoughlin 1969, p. 86.

The great value of the 'system' idea in the planning field was that it lifted the planning imagination from a narrow focus on the physical design of places, to a much richer emphasis on the dynamic socio-physical relations through which places were continuously evolving. However, these ideas were soon criticised for their assumptions that socio-economic systems were and should be inherently equilibrium-seeking. The structuralists argued that any apparent equilibrium was a temporary stabilisation in ongoing societal conflicts and usually meant that one group was imposing its values and interests on everyone else. But as social theory has evolved in the late twentieth century, an intellectual wave has developed which challenges the 'system' concept more fundamentally.

The traditional conception of the city as a system, while acknowledging the 'openness' of system relations, nevertheless assumed that internal relations were more important than external ones. Contemporary research on value-added chains and clusters in economic geography, on social networks and on the complexity of biospheric relations underlines not merely the openness of urban systems. It emphasises the diversity of the relations that transect urban areas, and the complexity and unevenness of their inter-relations (Graham and Healey 1999, Bridge and Watson 2000, Allen 1999). This implies not merely that cities are very open systems, but that the extent to which they exhibit 'systemness', in terms of the intersection of relations within a specific area, is open to question (see Figure 2.5).

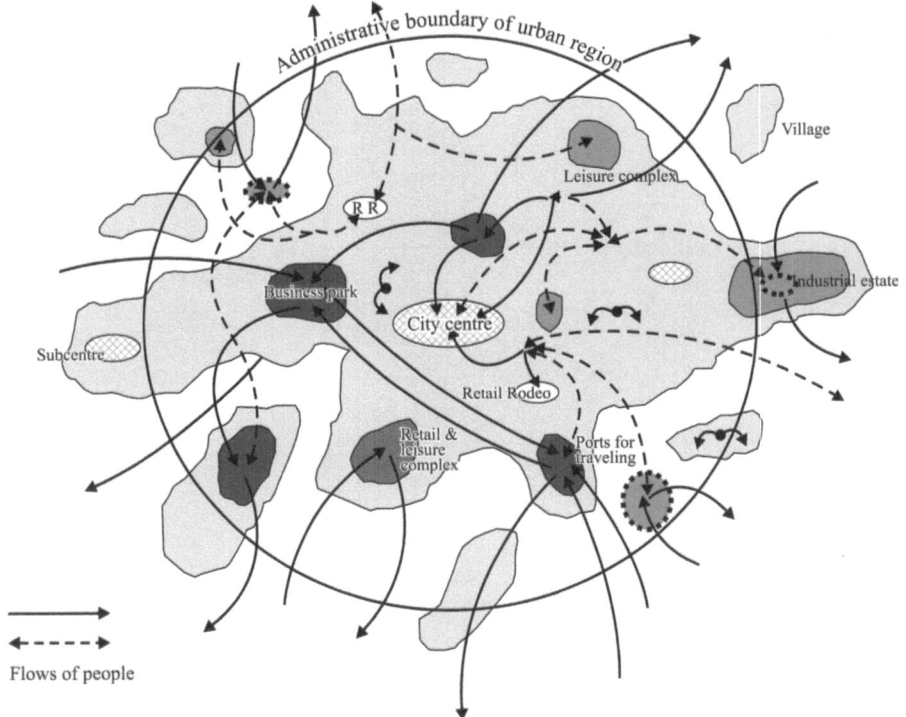

Flows of people

Figure 2.5   The multiplex city[10]

This conclusion has implications for the space-time of cities. The traditional city-as-a-system concept imagined a simple relation between activities and spaces. In planning texts of the 1960s, it was assumed that activities could be 'read off' and 'read into' land uses (Chapin 1965). Similarly relationships between places could be modelled using simple concepts of distance-decay, which assumed that activities near each other had more influence on each other than those further away. But contemporary geography introduces new concepts in socio-economic dynamics.

---

10 As devised by this author from ideas of Steve Graham, see Graham and Healey 1999, Healey 2000.

For example, New York and Tokyo are nearer to London than Paris, Pinner or Plumstead for those in the financial services industry. For those moving around Europe, hub airports and interchanges on Trans-European Networks provide the key nodes in their spatial maps while they leap over the spaces in between. Thus many different spatial scales and patterns transect the space of a city. These examples also suggest that multiple time horizons may be found in the urban context. There are not only multiple human timescales – a second, an hour, a day, a week, a year, a generation, but different natural timescales as well, from the lifecycle of a small insect, to timescales for water resource accumulation, and ultimately the time frame for geological formation.

The result of this openness and multiplicity of relational webs and of the complexity of their space-time dimensions is to undermine conceptions of the city as an 'integrated system', with a trajectory which can be 'guided'. Cities as such cannot be 'comprehensively' understood and planned for, because their dynamics are too complex. Contemporary planning theorists and policy analysts are turning to the analysis of 'complex systems' to provide different ways of imagining urban trajectories (Innes and Booher 1999, 2001). Drawing on 'complexity theory' developed in the physical sciences and climatology, they emphasise that, in highly open, complex systems, it is never possible to imagine all the relevant variables and how they interrelate. Yet even small actions can have effects, and be of a significant scale. Predicting a trajectory for the evolving city system and 'steering it' to maintain a given course are therefore impossible tasks. Not only will such an approach be 'torpedoed' by unexpected happenings from unimagined directions. It will be unable to see unexpected developments and emerging patterning. This raises all kinds of questions about the ability to identify the 'impacts' of actions, or even provide sets of general indicators with which to monitor the performance of systems and compare (benchmark) them with other similar systems (see Chapter 5).

The implication for strategic spatial planning is that all conscious interventions to shape urban dynamics need to be evaluated not as steps towards given ends, or movements along defined trajectories. Instead, they are risky experiments, taken within a particular hypothesis about what is important about urban dynamics and directed by particular values about what cities should be like. The conception of an 'urban system' is therefore not a mental construct that can be objectively tested against 'reality' for its robustness. It is tested against the many imagined realities brought into policy arenas by policy actors and legitimated through their diverse value systems. 'Objective reality' can only be known through our capacity to organise our observations and interpretations, and to give meaning to what we see. This however raises questions about the nature of knowledge and about the ways knowledge is mobilised in policy processes.

To summarise, the 'systems' referred to above are mental constructs, organising devices to find a way through the complexities of urban and regional relations and dynamics. In constructing them, selections are made as to critical nodes and relations, about what to integrate with what, and what 'whole' we see. The current force of the 'sustainable development' idea lies in its power to de-stabilise the 'system' models in use in policy communities. But if it simply replaces one model with another, it will remain trapped in traditional grooves. To maintain a radical edge, attention is needed to its epistemic dimensions.

2.6   The nature of knowledge

For much of the twentieth century, the dominant approach to knowledge production and use in public policy was framed within a positivist epistemology. This assumed a dualistic relation between 'observer' and 'reality'. By trained, scientific investigation of the 'objects' of the natural and social world, the principles by which these objects operated and related to each other could be extracted, through rigorous experimental and empirical testing. In this way, the principles could become 'objective laws'. Within planning and policy science in mid-twentieth century, some argued that, if these laws could be discovered, then they could be used to design policies to improve urban and regional environments (Faludi 1973). Contemporary philosophy of science, while recognising the power of scientific methods to 'discover' new dimensions of the dynamics of the 'world out there', is much more aware of the limitations on the human capacity to 'observe'. There is also much more attention to the way the frames of reference and interests which scientists, as well as everyone else, use to make sense of observations affect what is noticed and its interpretation. The various scientific communities can be understood as engaged in continual struggles over how to focus inquiry and how to interpret data and to give it meaning (Kuhn 1970, Barnes 1982, Latour 1987). Within these struggles, schools of thought, or epistemic communities, are as important as the experiments and data collected, since these frame the focus of scientific attention and the interpretation of data. Among scientists, these debates provide lively stimulants to further inquiry and interpretative effort. But they also flow into policy debate and general discussion where the contingent parameters and potential to revise any scientific conclusion is filtered out. Even in professional practices, where experts deploy specialised knowledge to solve problems for various kinds of clients, the debates over interpretation may be lost, with different members of a profession operating from different positions within the epistemic spectrum of a scientific field. As a result, in policy contexts, the capacity for 'objectiveness' of scientific inquiry, rather than acting as a stimulant to creative debate, tends to become a fuel for control over policy agendas between factions. This is very evident in Hajer's analysis of the acid rain controversy, and plays out, for example, in the debate on the Kyoto agreement and in the anguished discussion on the relative merits of culling and vaccination to control animal diseases.

Within planning theory and policy analysis, the shift away from a 'positivist' understanding of knowledge production was first interpreted in the language of political struggle. What counted as 'legitimate knowledge' was placed relative to the political status of proponents. The 'sustainability' discourses, with their fuzzy frames and claims, have value in the 'epistemic' dimensions of the struggles between and within policy communities because they draw in new frames of reference and give emphasis to new sets of data. But this dualistic conception of the relations between power and knowledge was itself undermined by the increasing recognition of the complexity of positions and power relations in play.[11] Urban and regional governance and policy dynamics in the US and Europe are increasingly described as 'diffused power situations', operating in a 'no-one in charge' world. As argued in Chapter 3, in such contexts, not only are there multiple positions among formal government actors, with their own epistemic frames and intellectual capital. All kinds of other actors get involved – various lobby groups, different businesses, as well as many citizens as individuals or in various kinds of groups. This introduces further complexity to the epistemic landscape. In addition to an array of scientific knowledge and professional understandings and practices, this widening of policy processes introduces all kinds of experiential knowledge, understood in emotive and aesthetic ways, derived from the many locales of life, both material and mental, of those contributing. The communicative/interpretative turn in planning theory and policy analysis emphasises the social dynamics of these encounters between different forms of knowledge and different frames of reference. Analysts emphasise not merely the power struggles between positions, but the social learning processes that evolve within such encounters, and how knowledge, understanding and new frames of reference are actually produced through them.

The result is a *social constructivist* approach to knowledge production (Hajer 1995, Hajer and Wagenaar 2002, Healey 1997, Innes 1990, Schon and Rein 1994). This recognises the impossibility of an objectively-grounded understanding of, say, urban and regional dynamics. Such an understanding has to be 'created' in some way from the knowledge resources available and legitimated by some socio-political process. The implications for the enterprise of strategic spatial planning are substantial.

If the power relations of urban regions are diffuse, then any strategy that is to be accepted as legitimate and effective must draw upon the multiple knowledge resources of the many actors whose 'power' will affect the rhetorical and material force of any strategy. This is not just a way of pooling information and testing understandings. It is also about creating new understandings that change the epistemic frames and knowledge resources deployed by actors in all kinds of situations. This places major demands on the design of policy processes – the selection of policy design arenas, choices about discursive practices and sequences, choices about who gets involved at what stage, and so on (Healey 1997, Innes 1992, Gualini 2001). If the knowledge resources are borrowed from somewhere else – from a consultants' tool kit, or from the manifesto of one of the participants, the legitimacy and effectiveness of any strategy is likely to be limited. It will not be 'owned' nor will it have 're-framed' the knowledge resources in use among the actors with a key role in the future of an area.

---

11 These are revealed very well in analyses of discursive struggles in European spatial policy and planning, see Richardson 1997, Jensen and Richardson 2000, Zonneveld 2000.

Critical to these processes are choices about the strategic focus of attention, and how linkages are made between different fields of concern. Whether to search for an 'integrated' approach, and what to integrate with what, is again not an issue that can be resolved by formula or general prescription. There is no 'objective' model of an urban or rural region robust enough to encompass the multiplicity of understandings of city and region to be found among both scientific and lay knowledge resources. As argued in Chapter 7, any 'model' with its nodes and links has to be socially constructed in situ to have meaning and power to transform practices. As with any model, it will serve to highlight some issues and situations, while others are relegated to the shade. The search for 'integration' and 'holistic models', such as the actor-consulting model presented in this book, may thus be understood as an effort to shift policy attention into new modes of articulation and synthesis, not as some grand mental synthesis.

The resultant social positioning of spatial strategies may give them power and legitimacy. The failure to pay adequate attention to this helps to account for the weakness of spatial strategies in the UK in recent years (Vigar et al. 2000). But such positioning raises big questions about the policy technology, which has built up in policy communities promoting 'sustainable development' in recent years. This is particularly so with respect to environmental impact assessment and sustainability indicators. These tend to assume that measures of impacts and conditions are somehow 'objective' measures of specific phenomena (see Chapter 5). Yet even with such measures as air quality, there are significant debates about the impact of different air quality conditions on health, about techniques of measurement, and about processes of measurement (Elsom 1999). Many other indicators and measures turn out to be even more contestable. The social-constructivist understanding of how knowledge is produced and used in policy contexts emphasises that these technologies need to be related to the context of their production and use. They should not be understood as objective measures that allow performance to be compared in space and time, but for example as devices to help actors monitor their behaviour and that of others in some way (Innes 2000). The Danish case study at Chapter 9 gives an example of the use of indicators as a means to promote debate. The design of impact assessment, indicators and monitoring processes therefore needs to be produced in relation to the role of the actors concerned, as explained in Chapter 5. The use and identification of indicators is thus a key part of the struggle to control, contain and direct the transformative potential of the 'sustainability' discourses.

2.7   The nature of policy processes

The above discussion of the nature of 'systemness' and the nature of knowledge implies that translating ideas about 'sustainable development' and 'sustainable cities' into policy frameworks and governance practices involves complex social processes in which the generation of concepts, meanings and knowledge are crucial dimensions. A 'system' does not exist to be discovered by good analysts. The meaning of 'sustainable development' is not a clear set of principles that can be applied once its proponents have captured control of key policy arenas. 'Systems' are created by imaginative endeavour and principles are evolved and applied in interactive processes through which knowledge, meanings and values are created. This implies that the 'sustainable development' idea needs to transform policy *processes* as well as policy frames, discourses and agendas.

It is this appreciation that lies behind the increasing emphasis in planning theory and policy analysis on the interactive and discursive dynamics of policy processes. This parallels the appreciation in politics that traditional forms of political organisation (in which a manifesto is created by a political party, which then captures control of the machinery of government to implement the manifesto) provide far too crude a policy process. This model fails to reflect the dynamic interplay of interest groups, public attitudes and corporate business influence, all of which may shape the policy agendas and practices. In a dispersed power world, 'command and control' models of policy-making and linear models of the relation of knowledge to action both founder.

It is in this context, combined with the increasing disenchantment of citizens in many countries with formal government and political processes, that a wave of enthusiasm for collaborative and deliberative forms of policy-making is sweeping across academic and policy arenas. These seem to offer opportunities for innovation and creativity in developing policy concepts and knowledge resources to support new ideas. They suggest mechanisms for building foundations of common understanding upon which new ideas can acquire legitimacy and shape the flow of dispersed action.

But it is increasingly clear that there is no one model of a 'discursive' or 'communicative' policy process (Woltjer 2000). In a general sense, all policy processes involve interactions of some kind and have a 'communicative' dimension. The current emphasis on the communicative dimensions of policy processes develops from an appreciation that the nature of these interactive and discursive processes matters more than the rationalists and structuralists thought (Hajer 1995, Innes 1995, Gualini 2001, Healey 1997). This implies that more attention needs to be given to:

- the arenas in which policy developments are articulated;
- who gets access to them;
- the nature of the knowledge and understandings deployed and created within them;
- the processes of translation and dissemination through which new policy ideas are diffused into governance practices and individual actions;
- the impact of policy processes on their wider legitimacy and accountability.

These approaches to the understanding and analysis of policy processes stress not only that policy agendas and policy relations co-evolve with each other. Process forms co-evolve, as agendas and relations evolve, beautifully illustrated in Gualini's account of the New Jersey 'cross-acceptance' process (Gualini 2001). They generate and make use of intellectual, social and political capital (Innes and Booher 1999, Healey 1990, 1998). The approach taken by the actor-consulting model emphasises the multiplicity of potential process forms, and the way in which processes are embedded in their institutional context.

This re-thinking of the nature of policy processes suggests that strategic spatial planning, should be understood as an enterprise in generating policy ideas with the power to frame ways of thinking and ways of acting over the long-term. This implies that such efforts should be judged not merely in terms of their formal powers or their policy contents. The critical evaluation should also focus on who and what is to be influenced and how such influence may be exercised. Such an evaluation should focus on the extent to which new intellectual, social and political 'capital' is generated and how far it challenges and transforms that embedded in established policy communities and provides resources for the emergence of new ones (Healey 1998, Cars et al. 2002). Efforts in strategic spatial planning are thus non-linear, interactive processes, situated in particular socio-political contexts and contingent on their setting. Through discursive processes, foci of attention are selected and frames of reference reinforced, shifted and transformed. Assessments of the transformative power of 'sustainability' discourses thus need to pay much more attention to the *dimensions of policy processes* than is to be found in the current calls for empowerment and public participation.

2.8   The idea of 'sustainability' and the transformation of spatial planning

There is now a widespread critique of the policy agendas and practices of the British planning system for its failure to articulate new strategic frames, which aim to combine 'economic competitiveness', 'environmental sustainability' and 'social cohesion', so often cited in EU, national and local statements on planning. The insertion of 'sustainable development' as an overarching policy objective in the UK has altered many emphases in the policy portfolio deployed through the plan-making machinery of the system. However, the evolution of the 1990s agenda has been paralleled by a sophistication of the planning system into a distinctive regulatory form. The key arenas for policy articulation remain the civil service offices responsible for developing the array of national-level policy criteria, the institutional establishments where these are interpreted into local plans and regulatory decisions, and where these are tested in inquiries and the courts. Recent efforts to breathe new life into the production of regional planning guidance and strategic plans through policy processes designed along broadly-based, collaborative lines, struggle against the embedded power of this semi-judicial process. This struggle weakens the imaginative, creative potential of collaborative processes, often leading to disillusionment among actors not previously involved in policy-making, as initial enthusiasm is blunted against the structure of the regulatory process. The persuasive power and legitimacy of both regulative and collaborative strategy-making processes are thus limited and continually challenged.

This is the containing power of established governance processes at work, blunting the transformative potential of 'sustainability' discourses.

This chapter argues that these discourses contribute to limiting their own transformative power unless some of the conceptual baggage carried with them is reviewed. It is suggested in particular that the promotion of 'sustainability' ideas need to connect to the debates on new ways of imagining cities and regions; articulating strategies; conceptualising the links between strategy and action; building governance capacity. The implications of a social-constructivist conception of such an approach have been explored.

Imagining cities is about articulating the linkages of policy frameworks, developing a *shared* awareness of the space-time dynamics of the relationships weaving through and across an area. It is about trying to see established and emerging socio-spatial patterns in these dynamics and imagining how they might play out in the future. It involves thinking creatively about actions that could make the future different to what seems to be emerging, or to encourage it along a trajectory that appears more desirable or 'sustainable' than current dynamics seem to suggest. The mode of imagining needs to be interactive, involving many stakeholders (those who have a vested interest in relevant social, economic and environmental issues) and actors (those whose actions or knowledge have the potential to make a difference) in the imagining process.[12]

Articulating strategies is not so much about the definition and specification of goals, as the generation of a way of thinking about issues, their connections, the array of actors involved, and what they might think and do. It is not about achieving a plan or vision with a 'comprehensive' coverage, which will inevitably end up partial and selective. It is about developing a broad angle of vision,[13] within which people can locate themselves. It should provide imaginative resources to support creative and innovative actions, which are shaped to avoid compromising what others want to do and care about. It should trigger new reactions, responses and understandings, and thus evolve in use, enriching locally available intellectual, social and political capital. It should avoid formulaic strategic concepts and regulatory principles.

Linking strategy and action involves moving away from hierarchical conceptions of policy control, as, for example, expressed in nationally prescribed regulatory rules and funding criteria. Instead, it takes the form of an interactive process with material and mental outcomes. This means not merely that attention needs to be given to how policy articulation relates to dissemination and translation. It also means that arenas need to be created for recursive flows between strategic articulation and operational situations.

These are demanding requirements. They are being achieved in some situations, where the inherited governance capacity provides fertile ground for generating the intellectual, social and political capital needed for such activity. But elsewhere, moving forward on giving transformative meaning to 'sustainable development'

---

12 Chapter 3 pursues a further discussion of the respective roles of 'citizens', 'stakeholders' and 'actors'.

13 This idea was developed in the 1960s by Etzioni (1973).

and 'sustainability' involves a political project to change the institutional capacity of urban and regional governance processes overall, so that new kinds of policy process can flourish. As will be seen in the following chapter, the *transformation of governance culture and process* then becomes the major project, rather than the transformation of specific policy agendas and practices. Ideas about 'sustainable development', 'sustainable cities', and so on, may contribute to this transformative project, but only if they can acquire meanings which have the power to mobilise stakeholders in many parts of the diffused urban and regional polity to create the capacity to act strategically for new shared agendas. Without such 'collective actor' capacity, and in the absence of techniques such as actor-consulting, the governance 'system' in cities and regions will be as fragmented and diffused as the webs of relations weaving across them, and discourses about 'sustainability' will have little transformative leverage.

References

Allen, J. (1999) 'Worlds within Cities', in: D. Massey, J. Allen and S. Pile (eds) *City Worlds*, Routledge/Open University, London, pp. 53–97.

Barnes, B. (1982) *T.S. Kuhn and Social Science*, Macmillan, London

Bertalanffy, L. von. (1970) *General Systems Theory*, Penguin, Harmondsworth (UK).

Bridge, G. and S. Watson (eds) (2000) *A Companion to the City*, Blackwell, Oxford.

Cars, G., P. Healey, A. Madanipour, and C. de Magalhaes (eds) (2002) *Urban Governance, Institutional Capacity and Social Milieux*, Ashgate, Aldershot (UK).

Chadwick, G. (1970) *A Systems View of Planning*, Pergamon Press, Oxford.

Chapin, F. S. (1965) *Urban Land Use Planning*, University of Urbana-Champagne Press, Urbana, Illinois (USA).

Committee for Spatial Development (1999) *The European Spatial Development Perspective*, European Commission, Luxembourg.

Counsell, D. (1999a) 'Attitudes to sustainable development in planning: policy integration, participation and Local Agenda 21, a case study of Hertfordshire Structure Plan', *Local Environment*, Vol. 4,(1), pp. 21–32.

Counsell, D. (1999b) 'Attitudes to sustainable development in the housing capacity debate: a case study of the West Sussex structure plan', *Town Planning Review*, Vol. 70, (2), pp. 213–230.

Department of the Environment (DoE) (1997) *Planning Policy Guidance Note 1: General Policy and Principles*, The Stationery Office, London.

ECOTEC Research and Consultancy (2000) The Urban Audit, http://www.inforegio. cec.eu.int/urban/audit/src/results.html.

European Commission (2001) 'Directive 2001/42/EC on the assessment of the effects of certain plans and programmes on the environment', *Official Journal of the EC*, 21.07.2001, Brussels.

Elsom, D. (1999) 'Development and Implementation of Strategic Frameworks for Air Quality Management in the UK and the European community', *Journal of Environmental Planning and Management*, Vol. 42, (1), pp. 103–122.

Etzioni, A. (1973) 'Mixed-scanning: a "third" approach to decision-making', in: Faludi, A (ed.) *A Reader in Planning Theory*, Pergamon, Oxford.

Faludi, A. (1973) *Planning Theory*, Pergamon Press, Oxford.

Forrester, J. (1968) *Urban Dynamics*, MIT Press, Boston.

Graham, S. and P. Healey (1999) 'Relational concepts in time and space: issues for planning theory and practice', *European Planning Studies*, Vol. 7, (5), pp. 623–646.

Gualini, E. (2001) *Planning and the Intelligence of Institutions*, Ashgate, Aldershot (UK).

Hajer, M. (1995) *The politics of environmental discourse*, Oxford University Press, Oxford.

Hajer, M. and H. Wagenaar (eds.) (2002) *Deliberative Policy Analysis: Understanding Governance in the Network Society*, Cambridge University Press, Cambridge.

Hall, P., R. Thomas, H. Gracey, and R. Drewett (1973) *The Containment of Urban England*, George, Allen and Unwin, London.

Haughton, G. (1999) 'Searching for the sustainable city: competing philosophical rationales and processes of "ideological capture" in Adelaide, South Australia', *Urban Studies*, Vol. 36, (11), pp. 1891–1906.

Healey, P. (1990), 'Policy processes in planning?' *Policy and Politics*, Vol.18 (1), pp 91–103.

Healey, P. (1997) *Collaborative planning: shaping places in fragmented societies*, Macmillan, London.

Healey, P. (1998) 'Building institutional capacity through collaborative approaches to urban planning', *Environment and Planning A*, Vol. 30, pp. 1531–1456.

Healey, P. (2000) 'Planning in relational time and space: responding to new urban realities', in: G. Bridge and S. Watson (eds) *A Companion to the City*, Blackwell, Oxford, pp. 517–530.

Healey, P. and T. Shaw (1994) 'The changing meanings of "environment" in the British Planning System', *Transactions of the Institute of British Geographers*, Vol. 19, (4), pp. 425–438.

Innes, J. (1990) *Knowledge and public policy: the search for meaningful indicators*, Transaction Books, New Brunswick (USA).

Innes, J. (1992) 'Group processes and the social construction of growth management', *Journal of the American Planning Association*, Vol. 58, (4), pp. 440–454.

Innes, J. (1995) 'Planning theory's emerging paradigm: communicative action and interactive practice', *Journal of Planning Education and Research*, Vol. 14, (4), pp. 183–189.

Innes, J. (2000) 'Indicators for sustainable development: a strategy building on complexity theory and distributed intelligence', *Planning Theory and Practice*, Vol.1, (2), pp. 173–186.

Innes, J. E., D. Booher (1999), 'Consensus-building and complex adaptive systems: a framework for evaluating collaborative planning', *Journal of the American Planning Association*, Vol. 65, (4), pp. 412–423.

Innes, J. and D. Booher (2001) 'Metropolitan development as a complex system: a new approach to sustainability', in: A. Madanipour, A. Hull and P. Healey (eds) *The Governance of Place*, Ashgate Aldershot (UK), pp. 239–264.

Jensen, O. and T. Richardson (2000) 'Discourses of mobility and polycentric development: a contested view of European spatial planning', *European Planning Studies*, Vol. 8,(4), pp. 503–520.

Kuhn, T.S (1970), *The Structure of Scientific Revolutions*, University of Chicago Press, Chicago.

Latour, B. (1987), *Science in action*, Harvard University Press, Cambridge, Massachusetts (USA).

Macnagthen, P. and J. Urry (1998) *Contested Natures*, Sage, London.

McLoughlin, J. B. (1969) *Urban and Regional Planning: A Systems Approach*, Faber and Faber, London.

Nadin, V., C. Brown and S. Duhr (2001) *Sustainability, Development and Spatial Planning in Europe*, Routledge, London.

Owens, S. (1997a) 'Giants in the path: planning, sustainability and environmental values', *Town Planning Review*, Vol. 68, (3), pp. 293–304.

Owens, S. (1997b) 'Interpreting sustainable development: the case of land use planning. Greening the Millennium?', in: M. Jacobs (ed.) *The New Politics of the Environment: Political Quarterly*, Blackwell, Oxford, pp. 87–97.

Ravetz, J. (2000) *City Region 2020: Integrated Planning for a Sustainable Environment*, Earthscan, London.

Richardson, T. (1997) 'The Trans-European transport network: environmental policy integration in the European Union', *European Urban and Regional Studies*, Vol. 4, (4), pp. 333–346.

Satterthwaite, D. (ed.) (1999) *The Earthscan Reader in Sustainable Cities*, Earthscan, London.

Sawicki, D. (2002) 'A proposal for improving community indicators: a modest research agenda and a temporary moratorium on practice', *Planning Theory and Practice*, Vol. 3, (1), pp. 13–32.

Schon, D., M. Rein (1994) *Frame reflection: towards the resolution of intractable policy controversies*, Basic Books, New York.

Selman, P. (2000) *Environmental Planning: the conservation and development of biophysical resources*, Sage, London.

Stewart, M. (2002) 'Compliance and Collaboration in Urban Governance', in: G. Cars, P. Healey, A. Madanipour and C. de Magalhaes (eds) *Urban Governance, Institutional Capacity and Social Milieux*, Ashgate, Aldershot (UK), pp. 149–167.

Vigar, G., P. Healey, A. Hull and S. Davoudi (2000) *Planning, governance and spatial strategy in Britain*, Macmillan, London.

Williams, K., E. Burton, M. Jenks (2000) *Achieving Sustainable Urban Form*, E&FN Spon, London.

Woltjer, J. (2000) *Consensus Planning: the relevance of communicative planning theory in Dutch infrastructure networks*, Ashgate, Aldershot (UK).

Wong, C. (2000) 'Indicators in use: challenges to urban and environmental planning', *Town Planning Review*, Vol. 71,(2), pp. 213–239.

Yanow, D. (1996) *How does a Policy Mean? Interpreting Policy and Organizational Actions*, Georgetown University Press, Washington, DC.

Zonneveld, W. (2000) 'Discursive aspects of strategic planning: a deconstruction of the "balanced competitiveness" concept in European Spatial Planning', in: W. Salet and A. Faludi (eds.), *The Revival of Strategic Spatial Planning*, KNAW, Amsterdam.

Chapter 3

Actors in a Fuzzy Governance Environment

Karel Martens[1]

3.1 Introduction

Governance processes differ widely between countries, localities and policy fields. They vary in their legal forms, patterns of resource flows, relations between governmental agencies, relations between public sector and private sector, and levels of community involvement. Even within a single locality governance processes may vary, depending on the issues at stake and the actors involved. The diversity in governance processes goes hand in hand with a wide variety in the roles and responsibilities of the actors involved in such processes. In some cases, a few actors may dominate much of the policy development and decision-making, while in other cases responsibilities may be fairly evenly divided over a large group of actors. Likewise, an actor may take the lead in one case, while the same actor may be on the fringes of a policy process in another case. Despite this variation, actors tend to base their expectations and actions on their experience of previous governance processes and on implicit frames of reference, often without questioning the role patterns that are embedded in these processes.

The goal of this chapter is to provide a framework for the wide variety of governance processes and the roles and responsibilities of the actors involved in them. It does so through the discussion of three prototype models of governance: the coordinative model, the competitive model and the argumentative model. The models differ in the roles, responsibilities and authorities they ascribe to the actors that are involved in and affected by governance processes. They are comparable in the sense that they provide a clear-cut frame of reference with regard to the role of various types of actor (§ 3.2). Many governance processes, especially in the field of spatial planning, have been based on the assumptions underpinning the coordinative model (§ 3.3). However, as we will go on to see, more and more governance processes integrate some of the elements of the competitive or communicative models into previously coordinative processes (§ 3.4). This chapter builds on the arguments presented in Chapter 2, which underlines the need to rethink 'the nature of policy processes', and in particular processes that are moving away from the coordinative or 'command

1   Karel Martens is Research Fellow at the Environmental Simulation Laboratory, Porter School of Environmental Studies, Tel Aviv University, Tel Aviv, Israel.

and control' models of governance. The straightforward models are being replaced by 'fuzzy' modes of governance, so called because the roles and responsibilities that planning authorities are expected to take are no longer straightforward as these are related to the activities of a wide range of parties including other then governmental bodies. Fuzzy modes have been part and parcel of past governance traditions, but recent developments have led to an increase in fuzzy governance modes coupled with an increasing variation in the roles and responsibilities of actors (§ 3.5). Actors need to be aware of this fluidity in the governance processes in which they operate (§ 3.6). The chapter concludes with some guidance on the shaping of modes of governance, relative to the role of the actors involved.

3.2   Three ideal models of governance

The variety in modes of governance is, as mentioned above, practically unlimited. The goal of this section is not to create order in this wide variety (see e.g. Mintzberg 1983, Harrison 1994, Huitema and Van Snellenberg 1997, Healey 1997, Coston 1998). The goal is, rather, to present three ideal models of governance that can serve as a yardstick to assess the attributes of real-life governance processes and the position of various types of actor in those processes. The three models – the coordinative model, the competitive model and the argumentative model – are briefly discussed below. The main focus is on the roles, responsibilities and authority that are ascribed in each model to the main actors involved in or affected by governance processes: political institutions, governmental agencies, private business interests, issue-oriented interest-groups, locality-based citizen groups, and 'ordinary' citizens.

Governance through coordination

The coordinative model has a long history in planning theory. It has its roots in notions of rationality, bureaucracy and systems theory (Friedman 1987: 87–136). The basic assumption of the model is the division between the governing body and the governed, or between government and society (Snellen 1987: 11). The governing body is positioned above the governed and has the task to steer society for the good of the governed. It is – in the ideal situation – operating as a single entity, which collects information, sets goals and priorities, and selects and implements policies. Coordination is the response to problems created by a governing body that is comprised of many departments, sections and factions. It is supposed to recreate unity in goals and policy measures through elaborate processes of mutual adjustment between the various factions (Mastop 1987: 293–299). The coordinative model thus defines governance processes as continuous efforts in coordination between the factions that build the governing body.

The division between the governing body and the governed creates a clear framework to determine the roles, responsibilities and authority of the various actors involved in policy development. The only actors that have authority to take decisions are part of the governing body. They are the ones that are able to articulate the public interest, to determine the need for intervention, and to select the 'best' policies and programmes. The system of representative democracy formalises the superior position of governmental bodies, as it creates a system in which the people as electors hand over the authority to the government as an institution guided by elected officials. The elected officials, in turn, hold the ultimate authority. Their task is to articulate the 'public interest' and to set priorities. The bureaucratic institutions of the government support the elected officials in developing policies and programmes and are responsible for the implementation of these plans and programmes once they are approved. The bureaucratic institutions thus act on behalf of the elected officials and it is from this that they derive their authority (Healey 1997: 220–221, Albrow 1997: 74–88). Coordination between these bureaucratic institutions or governmental agencies, in turn, is necessary in order to ensure that the policies approved by the elected officials and implemented by the governmental agencies further the same goals and support one another.

The role of actors other than governmental bodies is limited in the coordinative model. Bodies that belong to lower levels of government will in some cases be considered 'part of the family' and will participate fully in the efforts to coordinate governmental policy. In other cases, they will be viewed as one of the actors to be 'governed' and thus as part of the society. The roles and responsibility of these and other 'governed' actors are fairly limited in the coordinative model. At worst, they are perceived as objects that have to be steered. At best, they are considered to be suppliers of information to the governing body and as loyal followers of rules and policies of the government (Van Gunsteren 1976, Mastop 1987: 150–151, Teisman 1995: 24). The role of being steered is not only ascribed to the 'ordinary' citizen, but just as much to the 'ordinary' company, the 'ordinary' public transport operator, the 'ordinary' lower level government, and so forth. The role of information source is typically reserved for organisations that represent the interests of larger collectives, such as labour unions and major business organisations.

Governance through competition

The second model takes its inspiration from political theory, market economy, and the pluralist model of democracy (Friedman 1987: 160–161, Berveling 1994: 63–69, Teisman 1995: 25–29, Gertel and Law-Yone 1991: 175). In the competitive model governance is primarily seen as a competition between actors with diverse interests. Actors set goals and formulate policies independent from each other and try to achieve them through power struggles with competitors. The key mode that moves governance forward is the power resources of an actor. The more resources an actor has, the more he will be able to convince others of the benefits of its policies, the more he will be able to pressure others to accept its intentions, and the more he will be able to overcome (nimby-ist) protests. Cooperation with other actors will only occur if it suits both sides. It is neither a goal nor a prerequisite for governance activities. The metaphor of the market and its basic assumption that the pursuit of self-interest leads to the best outcome for the society as a whole, serves as the ideological base for the model (Teisman 1995: 25–29, Coston 1998: 365–366).

The model provides a rather simplistic frame of reference for the roles, responsibilities and authority of actors. As in market economics, all actors are autonomous and operate on a 'level playing field'. None of the actors is a priori positioned above the others. The consequence is that all actors play the same role: each has interests to pursue, and resources with which to maximise them (Healey 1998: 1534). This is true for national departments, local governmental bodies, private companies, professional interest groups, community organisations and even individual citizens. All these actors are perceived as operating based on narrowly defined interests. However, the theoretical 'level playing field' is disturbed by the uneven distribution of resources of power. Since 'power' is the modality that drives governance processes, the role of actors will depend on the powers they have. Powerful actors – like spending departments, departments with large discretionary powers, government companies with large budgets, and major businesses – will be able to develop their policies and implement their plans fairly easily and thus determine most of the actual governance activities. Weaker parties will struggle to actively promote and achieve their goals. Much of their activity will follow the actions and programmes of the powerful in an effort to make them reflect at least some of the weak party's interest. Many of the weaker national departments, smaller municipalities, small businesses, interest-groups, and community organisations will find themselves at the fringes of governance processes. Among the interest groups the politics of voice will dominate. Strong interest groups with substantial financial and professional resources will be able to make their voice heard and influence the policies of the powerful. Smaller interest groups with limited resources and low membership levels are much less likely to make their voice heard or to have an impact on the actions and programmes of the dominant actors. The 'ordinary' citizen, in turn, will hardly be able to influence governance processes, unless she/he works through existing organisations or creates new community organisation.

Like in the coordinative model, the political institutions act as a modus through which policies are adopted and implementation is approved. However, the role of the political institutions is not so much to articulate the public interest on behalf of society, as to arbitrate between the interests of the different groups and in this way to legitimise certain interests and policy proposals (Healey 1997: 222–223). The political institutions are, in other words, a source of authoritative power that all parties need in order to implement plans and programmes in a legitimate way (Berveling 1994: 68–69). This crucial position turns the political institutions into a target of fierce campaigns from all interested parties. Strong parties will try to convince the political institutions of the advantages of their plans and programmes in an effort to secure support for the plans and provide them with the necessary legitimacy. Weaker parties, in turn, will try to reveal the drawbacks, or even try to de-legitimise these plans and programmes.

Governance through argumentation

The third model takes its inspiration from the large body of literature on communicative planning (e.g. Healey 1995, 1997, Innes 1995, 1996a, Sager 1994, Forester 1999) and deliberative forms of democracy (e.g. Dryzek 1990, 1993, Bohman and Rehg 1997, Giddens 1994). The basic ideology that underlies this body of literature is the idea that governance should be a process of argumentation between all involved 'stakeholders'. Healey (1995: 49) refers to this process as 'inclusionary argumentation': 'public reasoning which accepts the contributions of all members of a political community and recognises the ways they have of knowing, valuing and giving meaning'. Innes and Gruber (1999) define it as collaboration: a process in which all stakeholders 'jointly search for actions and strategies'. The process of inclusionary argumentation thus creates a 'level playing field'. However, this 'level playing field' is hardly comparable to the one in the competitive model, as not power but reason dominates the processes of policy development and implementation. The ideal process of governance is devoid of all plays of power and solely dominated by the force of 'the good argument' (Dryzek 1990: 15). The objective of communicative governance processes is to 'pursue reasoned dialogue and consensus at each of the [...] discursive phases of development' (Fischer 1995: 20).

Communicative planning thought stresses the fundamental equality of all actors. These are often defined as 'stakeholders': actors with a 'stake' in an issue. The term stakeholders does not only encompass actors that have a stake because of the resources they manage or the formal positions they hold, but includes the whole universe of affected people (present and future, local and non-local) and the organisations through which they operate (Healey 1998: 1538). The definition of actors as stakeholders stresses the similarities between actors. In its extreme form, differences between governmental bodies, private sector agencies, interest groups and even 'ordinary citizens' become obsolete. Every actor is respected as an equal participant in the processes of inclusionary argumentation. The focus is on the knowledge, assumptions, arguments and solutions these actors bring to the table, rather than on the formal responsibilities, power resources and interests of the actors (Martens 2002).

The basic assumptions of the communicative mode of governance are at odds with the principles underlying the existing political institutions of representative democracy. The proponents of the communicative model emphasise self-governance of citizens or stakeholders, and stress the limitations of governance legitimised by representative democracy. They uphold that the existing political institutions and bureaucratic apparatus are incapable of defining the public interest. The public interest, they claim, is not pre-given but can only be constructed through a process of argumentation between stakeholders. The goal of such a process is 'the reconstruction of private or partial interests into publicly defensible norms' (Dryzek 1989: 110). In its extreme form, the argumentative model thus suggests that the ultimate source of authority is the collective of stakeholders engaged in a debate. This collective defines the public or shared interest, sets policies and programmes, and in principle even determines which role each of the stakeholders will fulfil in the implementation of the policies. The roles of various actors in governance processes are thus the result, rather than the starting point, of the argumentative debate.

The 'governance triangle'

The three models of governance described here are explicitly presented as ideal models. Each of them presents a clear-cut framework about the role of the actors involved in and affected by governance processes. Each model, too, depicts an extreme model of governance. The coordinative model is extreme because of the absolute division between state and non-state actors and because of the roles it ascribes to both types of actors. The competitive model is extreme in its celebration of power as the sole motor of governance and in its extreme form of individualistic, chaotic way of policy development and implementation (Miller, Hickson and Wilson 1999: 53–55). The communicative model, in turn, is extreme in its worship of argumentation as the prime motor for policy development and in its fundamental equality of all actors. Taken together, the three models demarcate the boundaries within which real-life governance processes can be positioned (see Figure 3.1).

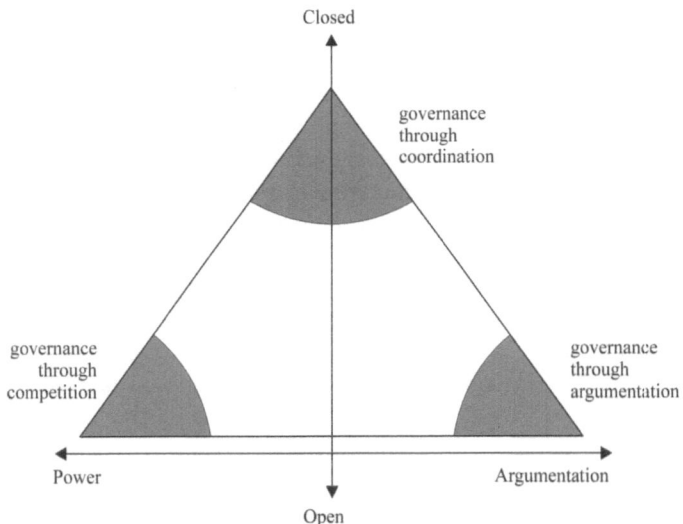

Figure 3.1   The 'governance triangle'

3.3   Dominance of the coordinative model

Generally, ideal models and reality do not show much overlap. It is different, however, for the coordinative model of governance. Much of the everyday governance practices of the past decades reflect many of the attributes of this model. This is especially true for modes of governance that dominated much of the fifties, sixties and seventies in European countries (Healey 1998: 1532–1533). The most extreme applications of the model are probably the efforts of central governments to develop systems for overall coordination of national policies and programmes. The Planning-Programming-Budgeting System in the United States and the Commission for the Development of Policy Analysis ('COBA') in the Netherlands are excellent examples of such efforts (Van Gunsteren 1976, Friedman 1987: 158, Mastop 1987: 150, 294).

The ideals and assumptions behind the coordinative model have not only left their imprint on everyday governance practices, but also found their way to formal institutional arrangements. This is especially true for the spatial planning field. Many countries have adopted formal planning systems that reflect the basic assumptions underlying the coordinative model. These systems give prevalence to governmental agencies in the preparation of policy programmes, request planning agencies to coordinate their efforts with related governmental bodies, consult major 'non' governmental organisations, and allow 'ordinary' citizens to voice their opinion in some sort of consultation process (Mastop 1996, Healey 1997). These systems thus formalised and reinforced the dominant role that governmental bodies played in everyday governance practices in the fifties and sixties. Moreover, they enabled planning agencies and especially the national-level planning apparatus to 'force' other actors into the role preferred by the governmental bodies. This has been especially true for the 'general public' or the 'ordinary' citizen (Robinson 2001: 83). The planning agencies have been much less successful in enforcing certain roles and positions on more powerful actors, like the private sector development industry, organisations of business interests, major companies, and large professionalised interest groups (see below).

The coordinative model still has a strong appeal among many governmental agencies and many governance processes still reflect the attributes of the model. Chapter 4 features a typical example from the Netherlands, based on the development of the Waterland regional plan. It shows how the Province of Noord-Holland used its formal authority to control policy development. The process of coordination and consultation was restricted to the relevant departments within the Province and to affected local governments. The role of other actors with an obvious stake in the plan, such as the agricultural sector, real estate developers and environmental groups, was limited to the legally prescribed forms of consultation. The plan that resulted from this process primarily serves as a tool for the Province to steer societal developments and 'governed' actors in the preferred direction.

3.4   Early fuzzy modes

The dominance of the coordinative model – reflected as much in actual governance practices and formal institutional arrangements as in the dominant discourse on governance – has never been absolute in the past decades. Many governance practices have reflected characteristics of more than one model of governance.

In many countries, for instance, processes of bureaucratic rivalry within large government organisations have introduced many competitive elements into the predominantly coordinative practices (e.g. Friedman 1987). In the US, the reliance on legal procedures and the dominant discourse of pluralist democracy has resulted in a politics of competing claims in which governance processes take the form of adversarial bargaining games (Innes et al. 1994). In countries like Germany and the Netherlands, the development of the corporatist model has introduced elements of the communicative model into governance processes (Healey 1997: 224–228). And in many countries around the world, public pressure has resulted in the – usually temporary and partial – dismantling of coordinative routines by participative practices that reflect some of the features of the communicative model (Figure 3.2).

The dominance of the coordinative model has thus gone hand in hand with governance practices that combine characteristics of more than one governance model. These practices are perceived as *fuzzy* in the sense that they are not based on a clear-cut framework about the roles and responsibilities of various actors. In the case of large bureaucracies, for instance, competitive tendencies lead to a fuzzy position for both governmental agencies and elected institutions. Rival governmental agencies will neither follow the formal coordination procedures, nor enter into an open competition with other agencies. Instead, they tend 'to live a life on their own' and define problems, set priorities, and decide about interventions in a relatively independent fashion. This, in turn, puts the elected institutions in an ambiguous position. They lose their ability to set overall priorities following the coordinative

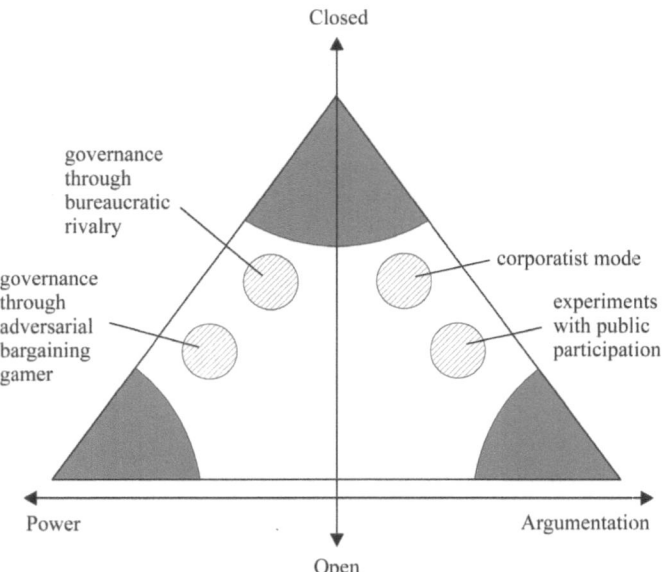

Figure 3.2   The position of early fuzzy modes within the 'governance triangle'

model, but also fail to acquire the position to arbitrate between conflicting interests as these conflicts remain inside the confines of the bureaucracy. The corporatist modes developed in the Netherlands and Germany display a comparable lack of clarity regarding the roles of governmental agencies and elected institutions. Corporatist arrangements are characterised by a situation in which a few powerful interests from inside and outside government determine the 'public interest' and define policies. The mode thus creates a situation in which both elected institutions and governmental agencies cannot fulfil the roles ascribed to them in the coordinative model. The elected institutions are robbed of their role to define the public interest and set priorities. The strong position of the governmental agencies that are part of the corporate alliance, in turn, obstructs the role of the bureaucracy as a whole to coordinate its policies.

3.5   The rise of fuzzy governance modes

While fuzzy modes of governance have thus been part and parcel of past governance practices, in recent years a sharp increase in the number of governance practices with fuzzy characteristics can be observed. More and more governmental bodies are experimenting with new modes of governance and the coordinative model seems to have lost some of its appeal. This can be attributed to a variety of reasons. Probably one of the most important is the legitimacy crisis of representative democracy. This crisis is strongly related to the rise of civil society and the growing diversity in lifestyles. The rise of civil society is visible in the increasing concern among citizens about the negative impacts of government policy (see for example Hajer and Kesselring 1999: 3). The growing diversity in lifestyles reflects the increasing number of social groupings defined around ethnicity, gender and life-style choices. This creates a growing distance between large parts of the electorate and elected officials, as the minority groups have a tendency to be under-represented (see e.g. Sandercock 1988: 122–125, Bloomfield et al. 2001: 507). Both the rise of civil society and the growing diversity of lifestyles have led to rising doubts about the ability of elected officials to define the public interest or to translate citizens' interests and preferences into policy (Edelenbos and Van Eeten 2001: 205, Beierle and Konisky 2001: 522). The legitimisation crisis of representative democracy thus erodes one of the crucial foundations of the coordinative model.

The legitimisation crisis of representative democracy is strengthened by the ongoing so-called regionalisation of local governance practices. (Chapter 4 further explores this phenomenon.) This process of regionalisation is a response to a rising awareness that problems cross the borders of local authorities and need to be addressed at a higher scale. The resulting governance practices necessarily include governmental bodies and other actors from both the local and supra-local levels. These processes thus pose challenges to the classical system of coordination, as they tend to focus on the intra-governmental dimension. Even more important is the fact that the regionalisation of governance challenges the link between democratic institutions and governmental bodies. The automatic link between the two is blurred in regional settings, as several representative bodies claim authority over the coordinative processes that affect their jurisdiction. Local representative bodies will tend to uphold that the higher level is not able to represent the interests of the local constituents, while the higher-level body is generally inclined to claim that the lower levels are unable to take care of the 'broader' public interests. The result of this struggle is often that regional governance practices are driven by strong governmental and quasi-governmental bodies without much democratic supervision (see Chapter 4). It is especially this democratic gap that has led to criticism on the adoption of the coordinative governance approach at the regional level (see e.g. Lukassen 1999: 48–51).

The rise of the notion of sustainability poses yet another challenge to the coordinative model of governance. The very radicality of the ideas underpinning sustainability questions the traditional organisation of government in sectors, hierarchies and levels and the established modes of governance that are related to them. But even when the idea of sustainable development is robbed of its radical edge and controlled by established governmental bodies, it still poses challenges to the coordinative model. Ironically, this is precisely because sustainability underscores the importance of a more comprehensive, holistic and integrated approach and thus pushes the need for coordination to its limits.

Finally, a more fundamental process undermining the coordinative model has to be mentioned: the increasing level of complexity of society (Dryzek 1990; Bloomfield et al. 2001: 501). Like high levels of specialization, high levels of complexity pose severe problems for coordination. This is not only the result of the need to 'connect everything with everything' in complex situations. More important is the fact that highly complex situations are very dynamic and dominated by non-linear processes and blurred relations between causes and effects, resulting in a fundamental level of uncertainty (De Roo 2002). This situation is at odds with the basic premise of the coordinative model, that the governing body is able to select effective and efficient interventions to steer society based on a thorough understanding of the causal relationships (Teisman 1995: 24–25). This approach fails in complex situations, simply because the coordinative model cannot deal with their dynamics and uncertainty. In complex situations knowledge cannot precede action, since knowledge only provides 'information about the likely effects rather than the exact effects of a proposed solution' (Jackson and Keys 1991: 143). Thus, knowledge and action have to be intertwined and governance has to be a process of trial and error, experimentation and adaptation to local and ever-changing situations (Innes and Booher 1997: 6). Comprehensive, top-down policies and programmes are thus doomed to fail in complex situations (Dryzek 1993; see also Chapter 2).

These and other developments have led to a rise in governance practices that can be dubbed 'fuzzy'. These have taken many forms. In some cases, more competitive modes of governance have been adopted as a response to the growing criticism on the coordinative model and government-dominated decision-making. In other cases public-private partnerships have been established in an effort to enhance governance capacities in an increasingly complex society. More recently, governance practices that adopt some of the features of the communicative model have started to mushroom especially in the last decade. Edelenbos and Monnikhof (2001: 9), for instance, show how participative modes of governance have 'invaded' local level governance practices in the Netherlands in the nineties, following the low levels of participation in local elections and the related crisis of local democracy. Greer (2001: 751) points out that partnerships – modes of governance that include public, private, voluntary and community sectors – have become 'a universal or global phenomenon'. And Bloomfield et al. (2001: 501) stress that 'opinion increasingly favours the use of deliberative and inclusionary processes as aids to decision making in public affairs'.

The actor-consulting model, which is at the core of this book, offers a means to cope with fuzzy modes of governance. It builds on both the coordinative and the communicative models of governance. On the one hand, the model relaxes the sharp division between governmental and non-governmental actors, which is so characteristic for the coordinative model. Non-governmental actors are not considered objects that have to be governed, nor solely as providers of information. Rather, they are perceived as active players in a policy field and as potential contributors to governmental policies. On the other hand, the actor-consulting model is all but an ideal example of the communicative model, being based neither on an extreme form of equality between stakeholders, nor on direct and extensive deliberations between these stakeholders. Rather, all actors reflect upon perceptions and interpretations of the roles they play (or believe they are playing) and the responsibilities they carry within a particular policy arena. The aim of the actor-consulting model is to come to a mutually agreed frame of reference that guides regulatory mechanisms in an efficient and effective way.

3.6 The need to address fuzziness

The rise of fuzzy modes is more than a sign of the times. It is a response to a highly dynamic society characterised by high levels of interconnectedness, high and diverging ambitions, and a myriad of actors that try to influence both procedures and content of governance processes. Under these circumstances it seems highly unlikely that a new, dominant, mode of governance will emerge. It is more likely that actors will be engaged in a 'restless search' (Offe 1977) for governance practices that match the issue and circumstances at hand and are perceived to be legitimate and effective by at least the stronger parties. These practices may be expected to be fuzzy and are likely to reflect elements of two or even three of the governance models presented in this chapter.

The fuzzy character of governance processes is in itself not a problem. The problem arises when this is not explicitly recognised, addressed and discussed among the relevant actors. Given the fact that the roles and responsibilities of actors are not ex ante given in a fuzzy process, a failure to address these issues may result in a lack of clarity among the involved actors about each other's roles and responsibilities. This, in turn, may have various negative effects that undermine the legitimacy and effectiveness of the governance process (see Figure 3.3).

The lack of clarity may first of all lead to *uncertainty* among actors regarding their role and responsibilities. This uncertainty can be observed among civil servants in governance processes that adopt argumentative characteristics. Many civil servants are used to working behind the protective walls of government bureaucracy, and have a strong orientation towards professional knowledge and expertise (Baum 1987). Governance processes that follow the rules of the coordinative model allow them to uphold this role. However, in governance processes that adopt some of the argumentative features, uncertainty will prevail if the role and responsibilities of civil servants is not redefined. This might at best induce civil servants to adopt a relatively passive role, and to follow a strategy of risk avoidance (Edelenbos et al. 2001: 222–224).

The lack of clarity may also result in *conflict* between the actors involved in a governance process. Actors from various policy fields and backgrounds will bring a wide variety of frames of reference into the process (Hillier 1998; Edelenbos 2000: 146–162), resulting in an increased probability that they will question the motives, preferences, ideals and actions of others. Given these diverging frames of reference, it is highly likely that an actor will base his actions on the role the actor ascribes to himself rather than on the roles others (implicitly) ascribe to him. The chances that certain actors will question the actions of other actors will thus be quite high. Greer (2001: 756–760) provides a good example in his analysis of the District Partnerships for Peace and Reconciliation in Northern Ireland. These partnerships brought together people from the public, private, and voluntary and community sectors, trade unions and local councils, people with widely diverging decision-making cultures and frames of reference. Tensions and conflicts were especially strong between representatives of the voluntary and community sectors and the local councils, as they based their expectations and actions on clearly different beliefs about democratic representation and legitimacy.

The lack of clarity regarding roles and responsibilities can lead to an *abuse* of roles by certain actors. Powerful actors may feel inclined to push their agenda forward 'behind the backs' of other actors. This tendency may be strong in governance models that adopt communicative and competitive characteristics, where powerful actors work both within and outside the framework of the fuzzy process in an effort to promote their interests (Hiller 1997; Neuman 2000: 345). Certain actors may also take advantage of the lack of clarity about roles and responsibilities and seize opportunities to set the agenda. Powerful actors may tend to abuse open planning processes, as they employ huge resources to promote their interests and thus marginalise the role of less organised interests (Bloomfield et al. 2001: 502; Hibbard and Lurie 2001: 191–192). Governmental agencies, in turn, often make use of the lack of clarity in the later phases of fuzzy processes, when the selection of policy options is at stake. In this stage, many governmental agencies take control of policy directions, largely outside the framework of the more open planning processes (Edelenbos and Van Eeten 2001: 208; De Bruijn et al. 2002: 16–17).

The lack of clarity in fuzzy governance modes can also heighten the risk of *stakeholder fatigue* in processes that adopt some of the characteristics of the communicative model (Healey 1998: 1534). Such processes tend to be demanding in terms of time, energy and intellectual capacity and thus tend to 'wear out' the actors participating in them. Small interest groups and 'ordinary' citizens will simply have difficulties to actively participate throughout the whole governance process. Larger, stronger interest groups, in turn, may also suffer from stakeholder fatigue when they feel that traditional modes of governance offer them more opportunities to pursue their agenda. Fuzzy governance processes that lack clarity about roles and responsibilities can therefore tend to aggravate these problems of stakeholder fatigue. They tend to create expectations among actors about their possible roles

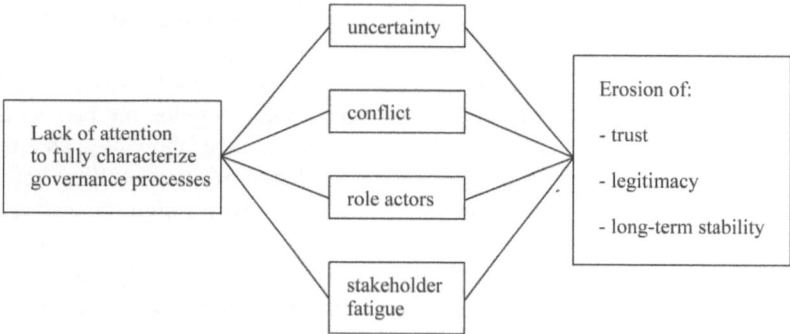

Figure 3.3   Possible consequences of a lack of attention for the fuzzy character of governance processes

and level of involvement, but fail to deliver the expected outcome (Bloomfield et al. 2001: 509).

When problems of uncertainty, conflict, role-abuse and stakeholder fatigue become severe, they are likely to damage the basis of the governance process. Conflict and role-abuse are especially prone to undermine the *trust* between the partners, the *legitimacy* of the governance arrangement, and the *long-term stability* of the process. Trust between partners is a crucial prerequisite for any governance process (Kumar and Paddison 2000). It allows participants to let down their guard and engage in policy processes without the constant need to question motives and monitor the actions of others (Beierle and Konisky 2001: 522). Legitimacy is especially crucial for fuzzy governance processes, as they depend more than well-established procedures on the support of all those involved. Conflict and role-abuse may induce actors to turn their backs on a fuzzy governance process and openly dispute its legitimacy. Stakeholder fatigue may have the same effect particularly in the case of communicative processes (Hillier 1998: 22). Long-term stability, in turn, builds on both trust and legitimacy and is crucial if fuzzy governance processes are to yield results in terms of approved policy decisions (Greer 2001: 754).

Many of the problems described above are likely to arise in virtually every fuzzy governance process that fails to address the roles and responsibilities of those involved, and that fails to address the fuzzy nature of the underlying concepts and notions. This makes fuzzy governance processes especially prone to criticism, as many actors actively participate in them and learn about their weaknesses and limitations. It may be expected, though, that fuzzy processes with more competitive features will suffer just as much. This is particularly true in North America, where many governance processes adopt a competitive form, as the various powerful actors strive to set the agenda and determine policy outcomes. However, these processes develop informally within the existing planning framework, and may fail to gain a de facto legitimacy. The result is that many disputes ultimately have to be solved in court (Healey 1997: 222–224).

The roles and responsibilities of actors thus need to be addressed in the case of most governance modes. It is not a realistic prospect however to define roles and responsibilities of actors before engaging in a fuzzy governance process. Practical experiences show that a set of new rules is not enough to guide the behaviour of stakeholders. Edelenbos et al. (2001: 223–224), for instance, show how an attempt to create equality among the actors *created* rather than prevented conflicts between the participants. They attribute this to the fact that actors with a high stake attempted to claim a stronger position.

While the definition of roles and responsibilities may be relatively easy on paper, changing the behaviour of actors in practice to adapt to a particular governance mode is a totally different thing. Actors will go their own way to pursue their individual goals, although some powerful government institutions may deliberately choose to adopt a particular role, or introduce a deliberate style of governance. Actors tend to base their behaviour on what they consider 'normal' and 'routine'. Even when they are formally expected to adopt a new role, this 'normal' behaviour will heavily influence the actual conduct of an actor. Especially in a highly dynamic context with a wide variety of governance processes, actors will simply not have time to internalise the roles and responsibilities ascribed to them in each governance process. There are no simple answers to solve this problem. The most promising way out here is probably a shift from a 'control' approach to an 'adaptive' approach (Edelenbos 2000; Edelenbos et al. 2001; De Bruijn et al. 2002). Such an adaptive approach would move away from the tendency to control new, fuzzy governance processes through a clear, pre-determined set of rules and a fixed process structure. It would instead focus on the way in which a process develops over time, on the types of roles and responsibilities actors actually adopt, and on the type of interaction culture that evolves. These would then be used as the input for a regular adjustment of the rules and process structure of the fuzzy governance process.

3.7 Influencing the nature of fuzzy governance modes

The recognition of fuzzy modes in current governance processes poses substantial challenges for all actors involved in these processes. No longer is it possible to rely on pre-defined procedures and responsibilities. In many cases, actors may rely on fuzzy modes of governance that have proven themselves in the past or in other policy fields. In other cases, traditional coordinative processes may be able to tackle the issues at hand. However, in a number of cases it may be advantageous to develop tailor-made governance processes from the start. The observation that the coordinative model can no longer be the 'default' mode of governance does imply however that a more explicit process of governance mode selection is necessary in order to achieve results.

Governmental agencies will no doubt play a prime role in promoting a suitable governance mode for certain planning issues. Under traditional institutional arrangements they are the only actors that have the ability to grant legitimacy to new modes of governance. Other actors are generally in a weak position to propose alternatives, or to dispute the legitimacy of existing or proposed modes. Similarly, they generally do not have the *power* to shape the character of new governance modes. Martens (2001), for instance, shows that even in cases where citizen groups and non-governmental organisations force governmental bodies to change existing governance procedures, it is ultimately the governmental agencies that shape the new processes. Thus, while government agencies may involve other actors in the selection, development and design of a new governance mode, they hold the keys to legitimise the new mode, to define its relation to formal planning procedures, and to link it to crucial centres of power (Neuman 2000: 345).

Governmental agencies thus play a key role in the selection and development of fuzzy modes of governance. This raises the question as to how these agencies can select a relevant mode of governance. We will therefore go on to discuss some criteria that may guide governmental agencies in the selection of a governance mode that fits the relevant circumstances, and that matches the demands and expectations of the actors involved. The first two criteria relate to the content of a policy issue, the latter two relate to the institutional setting.

Complexity of the planning issue

The complexity of a planning issue is the first characteristic that should be taken into account in the selection and development of a fuzzy governance mode. De Roo (2003: 145–151) shows that the mode of governance should be directly related to the level of complexity of the planning issue. Approaches that are close to the coordinative model are still valid for relatively simple planning issues. Such issues are characterised by clarity and agreement about policy objectives and clear relationships between causes and effects, which allows governmental bodies to adopt more centralised and hierarchical approaches. Very complex planning issues, in turn, require more communicative modes of operation. They are characterised by non-linear processes, blurred relationships between causes and effects, and continuous changes. . These characteristics require a governance approach that is able to deal with the dynamics and uncertainty of a planning issue. More communicative approaches are a feasible answer here, as they allow for processes of trial and error, experimentation and adaptation to ever-changing situations.

Woltjer (2000: 237–240) also underlines the importance of complexity in the selection of governance processes. He distinguishes four types of problems and four related modes of governance. Like De Roo, he reserves coordinative approaches for straightforward, technical problems that are characterised by consensus about goals and products. Distributional problems are more complex, as they are characterised by disagreement about the distribution of benefits. Here, negotiation processes in which stakeholders trade interests and compensation are more suitable. Sensitive problems are yet another grade more complex, as they are characterised by conflicts over goals and values. Here, will-shaping processes between dominant (government) actors could be a viable governance approach. Finally, Woltjer follows De Roo's suggestion that very complex problems require a governance approach that adopts its main characteristics from the communicative model. Woltjer talks here about a learning process aimed at mutual understanding between all involved stakeholders. In Figure 3.4, the four governance modes are positioned within the governance triangle.

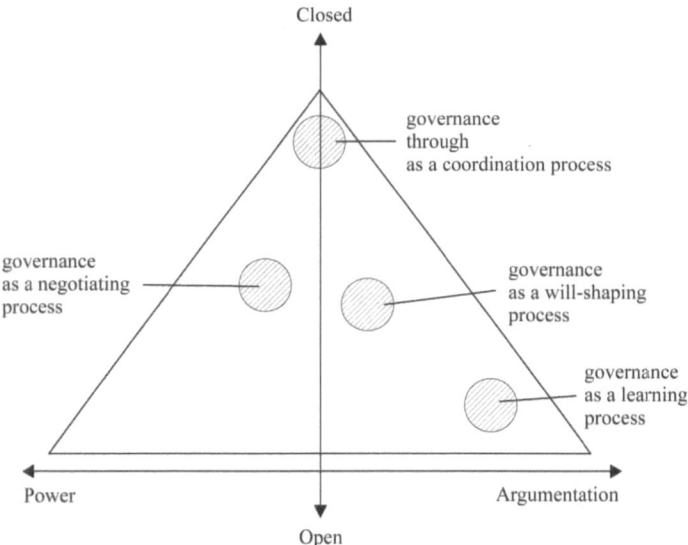

Figure 3.4   The position of four governance modes distinguished by Woltjer (2000) within the governance triangle

Appeal to citizens

A second factor that should play a role in the selection and development of a governance mode is the 'appeal to citizens' of a specific issue, reflecting the extent to which citizens may feel the inclination to become actively involved. Various authors have documented the problem of building a truly participative process (e.g. Woltjer 2000; Hibbard and Lurie 2001). Many participation programmes in Western Europe and the US tend to attract people that are also involved in the traditional consultation processes of the coordinative model: mature, white, highly educated and politically interested males (Edelenbos and Van Eeten 2001: 207). It has proven to be very difficult to attract a 'representative' selection of the population, or to attract minority groups. The issue at stake plays an important role here. Edelenbos et al. (2001), for instance, show that citizens are not very eager to participate in processes that deal with abstract, complex issues or issues with unclear impacts. In contrast, issues with clear direct impacts and issues that are sensitive in the local community tend to appeal to 'ordinary' citizens. The 'sex-appeal' of an issue should thus be considered in the selection and development of a governance mode. 'Heavy' participative modes of governance – modes that can be positioned in the bottom-right corner of the 'governance triangle' – are probably more likely to succeed around policy issues that affect citizens directly. More abstract and complex issues, in turn, may require a governance approach that balances coordinative and communicative approaches. In such cases, issue representation may be more effective than full stakeholder and citizen representation (Beierle and Konisky 2001: 517). Subsequently, the actor-consulting approach addressed in this book (Chapter 7) may be a very viable model here.

Institutional inertia

The term inertia refers to the ease with which an institutional setting can be changed. Institutional settings that are dominated by long established routines and procedures tend to be highly inert and difficult to change. Institutional arenas that are characterised by a diversity of working routines rather than one dominant mode of operation tend to be less inert and can be changed more easily (Dryzek 1990). The level of inertia will thus influence the range of governance modes that can be adopted. In more 'inert' institutional contexts the range of possible governance modes that can be considered will be limited, as actors will tend to oppose drastic changes into the dominant routines. Most likely only small adjustments to the dominant routine will be deemed acceptable. In contrast, in less 'inert' contexts it will be possible to consider governance modes that differ more substantially from the dominant or ordinary routines in the policy arena. Furthermore, it may be expected that the rise of fuzzy governance modes will go hand in hand with a reduction in the level of inertia in a policy field, as actors will get used to more regular changes in roles, responsibilities and various levels of stakeholder involvement. This suggests that the rise of fuzzy modes of governance will increasingly allow governments to choose between an increasing variance of governance modes (De Bruijn and Heuvelhof 1995). From the perspective of governmental agencies it will thus become increasingly important to identify current and possible roles of actors and to understand their perceptions and motivations. The actor-consulting model may have a role to play in this context.

Institutional complexity

The last criteria to be discussed here is institutional complexity. The number of actors in a policy arena, the relationships between them, the distribution of power and resources, the level of regulation and formalization, the level of interdependency, the culture of decision making, and the various frames of reference, all influence the complexity of a policy field. Institutional complexity is a function of the interaction between these features. For instance, a limited number of actors is obviously related to a low level of complexity, but a setting with a huge number of *similar* actors may in practice be characterised by the same level of complexity. Likewise, a clear concentration of power in the hands of one actor may be just as simple or complex as an even distribution of powers in a level playing field. While the assessment of the complexity of an institutional setting thus requires a thoughtful approach, it is an important criterion in the selection of an adequate governance mode. Much of the recent planning literature tends to stress the similarities between current institutional settings and describes them in terms of multi-actor constellations, mutual dependency, and shared power (see e.g. Bryson and Crosby 1993; Teisman 1995). Based on this observation, many authors propose various communicative approaches as a way to deal with the perceived complexity. However, a closer look at the wide variety of actual institutional settings may reveal that many policy issues are dealt with in relatively simple institutional settings. Governance in a more simple institutional setting may not rest in the hands of a single actor. However, the type of actors that are involved will be limited and/or relatively comparable in terms of policy culture and frames of reference. Also, contrary to the popular assumption of 'mutual' interdependency between each and every actor, real dependency in terms of e.g. blockage power may in many cases be limited to a few dominant actors. In such cases, more coordinative approaches to governance may be well suited to deal with the policy issues at hand. Also, the actor-consulting model seems to be suited to less complicated institutional settings. More complex settings, in turn, seem indeed more suited to communicative modes of governance (Healey 1997).

3.8   Finally...

This chapter builds on Chapter 2, by addressing the need to work towards a common frame of reference between the various parties that are gathered around a particular planning issue. Here we focus on the changing, integrating, and consequently fuzzy institutional context in which the inter-relationship between actors and their respective frames of reference are at stake. These frames of reference, and therefore the interpretation of the notions and concepts used in the planning arena, are role-dependent. In other words, the way in which an actor thinks and acts about an issue depends on his role in the planning process. The role of each actor will affect, and be affected by, the mode of governance in place for a particular planning issue.

This chapter has addressed some of the criteria that might come into play in determining the structure of a fuzzy mode of governance. The selection and development of a governance mode is normally in the hands of a government agency or a planning authority. In trying to influence the mode of governance, an analysis of the role, motivation and preferences of the various actors will be needed. Planning authorities will need to develop the capacity to adapt to new and ever changing governance modes. Only if such a capacity is built up, will actors be able to operate successfully in a fuzzy governance environment.

The actor-consulting model presented in this book can assist governmental agencies and others in answering these challenges, as it helps to define roles and responsibilities, and as a consequence, may eliminate certain aspects of fuzziness.. Because clarifying the various roles of actors within the planning arena is not in itself enough, the actor-consulting model will also need to address contextual power relations among actors, which can be achieved by examining the actual and desired behaviour of actors. The outcome should be a mutually agreed regulatory framework that guides all actions in an efficient and effective way. This provides actors with a basic understanding that can help them to cope with fuzzy governance processes.

In the following chapter, we will examine a particular level within the governmental planning system: the regional level of governance. This level sits between the national and the local levels, and displays qualities of a particularly ambiguous nature...

References

Albrow, M. (1997) *Do organisations have feelings?*, Routledge, New York.

Baum, H.S. (1987) *The invisible bureaucracy: the unconscious in organisational problem solving*, Oxford University Press, New York.

Beierle T.C., D.M. Konisky (2001) 'What are we gaining from stakeholder involvement? Observations from environmental planning in the Great Lakes', *Environment and Planning C: Government and Policy*, Vol. 19, pp. 515–527.

Berveling, J. (1994) 'Het stempel op de besluitvorming: macht, invloed en besluitvorming op twee Amsterdamse beleidsterreinen' [Marks in decision-making: power, influence and decision-making of two sectors of Amsterdam policy], Dissertation Rijksuniversiteit Groningen, Thela Thesis, Amsterdam.

Bloomfield, D., K. Collins, C. Fry and R. Munton (2001) 'Deliberation and inclusion: vehicles for increasing trust in UK public governance?', *Environment and Planning C: Government and Policy*, Vol. 19, pp. 501–513.

Bohman, J. and W. Regh (1997) *Deliberative democracy: essays on reason and politics*, Massachusetts Institute of Technology, Cambridge (US).

Bryson, J.M. and B.C. Crosby (1993) 'Policy planning and the design of forums, arenas, and courts', *Environment and Planning B: Planning and Design*, Vol. 20: pp. 175–194.

Coston, J.M. (1998) 'A model and typology of government-NGO relationships', *Nonprofit and Voluntary Sector Quarterly*, Vol. 27(3), pp. 358–382.

De Bruijn, H. W. and Dicke H. van der Voort (2002) 'Procesmanagement en politiek-ambtelijke verhoudingen: dilemma's en strategieën van ambtenaren' [Process management and politics-administrative relations: dilemmas and strategies of government officials], TU Delft, Faculteit Techniek Bestuur en Management, Delft (NL).

De Bruijn, J.A., E.F. ten Heuvelhof (1995) *Netwerkmanagement: strategieëen, instrumenten en normen* [Network management: strategies, instruments and values], Lemma, Utrecht (NL).

De Roo, G. (2002) 'Coping with change, uncertainty and a growing complexity of our social and physical environment: adaptive responses and new concepts to tackle difficulties in the physical planning of the Netherlands', Paper presented at the XVI Aesop Congress in Volos, Greece, 10–15 July 2002, Faculty of Spatial Sciences, University of Groningen, Groningen (NL).

De Roo, G. (2003) *Environmental Planning in the Netherlands: Too Good to be True; From command and control planning towards shared governance*, Ashgate, Aldershot (UK).

Dryzek, J.S. (1989) 'Policy sciences of democracy', *Policy*, Vol. 22(1), pp. 97–118.

Dryzek, J.S. (1990) *Discursive democracy: politics, policy and political science*, Cambridge University Press, Cambridge (UK).

Dryzek, J. (1993) 'Policy analysis and planning: from science to argument', in: F. Fischer and J. Forester (eds) *The argumentative turn in policy and planning*, Duke University Press, Durham (UK), pp. 213–232.

Edelenbos, J. (2000) 'Proces in vorm: procesbegeleiding van interactieve beleidsvorming over lokale ruimtelijke projecten' [Process in form: process support of interactive policy-making as part of local spatial projects], Dissertation Technische Universiteit Delft, Lemma, Utrecht (NL).

Edelenbos, J. and M. van Eeten (2001) 'The missing link: processing variation in dialogical evaluation', *Evaluation*, Vol. 7(2), pp. 204–210.

Edelenbos, J. and R. Monnikhof (2001) 'Opzet en hoofdvragen van het onderzoek', in: J. Edelenbos and R. Monnikhof (eds) *Lokale interactieve beleidsvorming: een vergelijkend onderzoek naar de consequenties van interactieve beleidsvorming voor het functioneren van de locale democratie* [Local interactive policy-making: A comparative research for the consequences of interactive policy-making in relation to local democracy performance], Lemma, Utrecht (NL), pp. 9–14.

Edelenbos, J., R. Monnikhof, J.W. Duyvendak, A. Edwards, I. Horstik, J. Koppenjan, A. Krouwel, R. van de Peppel and A. Wilde (2001) 'Interactieve beleidsvorming: de inspraak achterna?' [Interactive policy-making: the participation behind?] in: J. Edelenbos and R. Monnikhof (eds) *Lokale interactieve beleidsvorming: een vergelijkend onderzoek naar de consequenties van interactieve beleidsvorming voor het functioneren van de locale democratie* [Local interactive policy-making: a comparative research for the consequences of interactive policy-making in relation to local democracy performance], Lemma, Utrecht (NL), pp. 215–232.

Fischer, F. (1995) *Evaluating public policy*, Nelson Hall Publishers, Chicago.

Forester, J. (1999) *The reflexive practitioner: encouraging participatory planning processes*, MIT Press, Cambridge/London.

Friedmann, J. (1987) *Planning in the public domain; From knowledge to action*, Princeton University Press, Princeton, New Jersey (US).

Gertel, S., H. Law Yone (1991) 'Participation ideologies in Israeli planning', *Environment and Planning C: Government and Policy*, Vol. 9, pp. 173–188.

Giddens, A. (1994) *Beyond left and right*, Polity Press, Cambridge (UK).

Greer, J. (2001) 'Whither partnership governance in Northern Ireland?', *Environment and Planning C: Government and Policy*, Vol. 19, pp. 751–770.

Gunsteren, H.R. van (1976) *The quest for control*, John Wiley, London.

Hajer, M.A. and S. Kesselring (1999) 'Democracy in the risk society? Learning from the new politics of mobility in Munich', *Environmental Politics*, Vol. 8(3), pp. 1–23.

Harrison, K. (1995) 'Is cooperation the answer? Canadian environmental enforcement in comparative context', *Journal of Policy Analysis and Management*, Vol. 14(2), pp. 221–244.

Healey, P. (1995) 'Discourses of integration: making frameworks for democratic urban planning', in: P. Healey, S. Cameron, S. Davoudi, S. Graham, and A. Madanipour (eds) *Managing cities: the new urban context*, John Wiley and Sons, Chichester (UK), pp. 251–272.

Healey, P. (1997) *Collaborative planning: shaping places in fragmented societies*, Macmillan Press, London/Hong Kong.

Healey, P. (1998) 'Building institutional capacity through collaborative approaches to urban planning', *Environment and Planning A*, Vol. 30, pp. 1531–1546.

Hibbard, M., S. Lurie (2001) 'Saving land but losing ground: challenges to community planning in the era of participation', *Journal of Planning Education and Research*, Vol. 20, pp. 187–195.

Hillier, J. (1998) 'Beyond confused noise: ideas toward communicative procedural justice', *Journal of Planning Education and Research*, Vol. 18, pp. 14–24.

Huitema, D. and A.H.L.M. van Snellenberg (1997) 'Beleid in stijl' [Policy in style], *Beleidwetenschap*, Vol. 11(1), pp. 55–72.

Innes, J.E. (1995) 'Planning theory's emerging paradigm: communicative action and interactive practice', *Journal of Planning Education and Research*, Vol. 14(3), pp. 183–190.

Innes, J.E. (1996) 'Planning through consensus building: a new view of the comprehensive ideal', *Journal of the American Planning Association*, Vol. 62(4), pp. 460–472.

Innes, J.E. and D.E. Booher (1999) 'Consensus building and complex adaptive systems: a framework for evaluating collaborative planning', *Journal of the American Planning Association,* Vol. 65(4), pp. 412–423.

Innes, J.E. and J. Gruber (1999) 'Planning strategies in conflict: the case of regional transportation planning in the bay area', Paper for the XIII AESOP Conference, 7–11 July 1999, Bergen, Norway.

Innes, J.E., J. Gruber, M. Neuman and R. Thompson (1994) 'Coordinating growth and environmental management through consensus building', California Policy Seminar, University of California, Berkeley (US).

Jackson, M.C. and P. Keys (1991) 'Towards a system of systems methodologies', in: R.L. Flood and M.C. Jackson (eds) *Critical systems thinking*, John Wiley and Sons, Chichester (UK), pp. 140–158.

Kumar, A. and R. Paddison (2000) 'Trust and collaborative planning theory: the case of the Scottish planning system', *International Planning Studies*, Vol. 5(2), pp. 205–233.

Lukassen, U.J.B. (1999) 'Fragmentatie en eenheid in ruimtelijk beleid: het netwerk van furo's en provinciale eenheden in het KAN-gebied' [Fragmentation and unity in spatial policy: network of 'furo's' and provincial units in the KAN-region], Dissertation Katholieke Universiteit Nijmegen, Nijmegen (NL).

Martens, K. (2001) 'The bright side of power: NGOs struggling for more participative decision-making', Paper for the Annual Conference of the Israeli Geographical Association, 11–13 December 2001, Tel Aviv, Israel.

Martens, K. (forthcoming) 'Participatory decision-making and sustainability: the role of environmental organisations', in: D. Miller and E. Feitelson (eds) *Planning for sustainable development in Israel*, Ashgate, Aldershot (UK).

Mastop, J.M. (1987) 'Besluitvorming, handelen en normeren: een methodologische studie naar aanleiding van het streekplanwerk' [Decision-making, acting and valuing: a methodological study in relation to regional plans], *Planologische Studies 4*, Universiteit van Amsterdam, Amsterdam.

Mastop, J.M. (1996) 'Dutch national planning at the turning point: re-thinking institutional arrangements', Paper for the International Seminar on National-level Planning Institutions and Decisions, 7–9 January 1996, Haifa, Israel.

Miller, S.J., D.J. Hickson and D.C. Wilson (1999) 'Decision-making in organisations', in: S.R. Clegg, C. Hardy and W.R. Nord (eds) *Managing organisations: current issues*, Sage, London/Thousand Oaks/New Dehli.

Mintzberg, H. (1983) *Structure in five: designing effective organisations*, Prentice-Hall, Englewood Cliffs.

Neuman, M. (2000) 'Communicate this! Does consensus lead to advocacy and pluralism?', *Journal of Planning Education and Research*, Vol. 19, pp. 343–350.

Offe, C. (1977) 'The theory of the capitalist state and the problem of policy formation', in: L.N. Lindberg and A. Alford (eds) *Stress and contradiction in modern capatalism*, Heath, Lexington (US), pp. 125–144.

Robinson, A. (2001) 'Framing Corkerhill: identity, agency and injustice', *Environment and Planning D: Society and Space*, Vol. 19, pp. 81–101.

Sager, T. (1994) *Communicative planning theory*, Avebury, Aldershot (UK).

Sandercock, L. (1998) *Towards cosmopolis: planning for multicultural cities*, Wiley, New York.

Snellen, I.Th.M. (1987) *Boeiend en geboeid: ambities en ambivalenties in de bestuurskunde* [Fascinating and fascinated by: ambitions and ambivalences in the science of public adminstration], Samsom, Alphen aan den Rijn (NL).

Teisman, G.R. (1995) *Complexe besluitvorming: een pluricentrisch perspectief op besluitvorming over ruimtelijke investeringen* [Complex decision-making: a pluricentric perspective on decision-making of spatial investments], Second edition, Vuga, The Hague (NL).

Woltjer, J. (2000) *Consensus planning: the relevance of communicative planning theory in Dutch infrastructure development*, Ashgate, Aldershot, (UK).

Chapter 4

From Government to Governance: Actor Participation in Regional Planning

Henk Voogd and Johan Woltjer[1]

4.1 Introduction

It is perhaps fair to say that throughout Europe, the numerous benefits of regional planning have not been fully exploited (Balchin et al. 1999). In many ways this is understandable. In most European countries regional planning is still taking place within a coordinative mode of governance, a rather classical form of governance as we have seen in Chapter 3, and therefore relatively abstract by nature, and perceived by citizens as having little direct relevance to their day-to-day concerns. Media attention consequently tends to gravitate towards national agendas or to the conflicts surrounding local planning issues. The public may therefore be forgiven for thinking of regional planning as an unnecessary tier of public administration, and an unworthy recipient of public funding. Nevertheless, regional planning has many *potential contributions* (see Chapter 8) to offer towards – for example – mediating between conflicting frames of reference at the local level, resolving social dilemmas (see § 4.3), and taking a lead in tackling issues that exceed the local level of planning. In particular this last example is becoming increasingly important, for various reasons. These include firstly a shift from a coordinative mode of governance towards more fuzzy modes of governance, and secondly the growing importance of subsidiarity, presenting the idea that issues could best be solved at the lowest competent level of planning. This gradual change in planning responsibilities complicates the various roles of actors at the regional level. Therefore the role of actors and the opportunities for actor consultation in regional planning is the subject matter of this chapter. Although the examples referred to here are from the Netherlands, the principles apply more generally to planning throughout Europe and to other parts of the world.

'Regional' planning tends to take on a different meaning in the context of each national policy arena, so for introductory purposes, we must attempt to define regional planning in a generic sense. Four basic roles for planning authorities acting at the regional level are identified here. Firstly, regional authorities occupy an intermediate level between the local and the national authorities. Secondly, regional authorities

---

1   Henk Voogd is Professor in Planning at the University of Groningen, Faculty of Spatial Sciences. Johan Woltjer is Assistant Professor in Planning at the University of Groningen, Department of Spatial Planning and Environment, Faculty of Spatial Sciences.

will set guidelines for local policy making. Thirdly, they will take the lead in solving planning problems that are typified by competition or even conflict between local authorities. And fourthly, depending on the extent of their political power, regional authorities will have their own particular social, economic and environmental agendas that they wish to pursue. While the public might be forgiven for not seeing the benefits of regional planning, it is also understandable that planners, policy-makers, politicians and other stakeholders may see regional planning as a fuzzy, unclear process, partly caused by regional planners having to play various roles at the same time.

This chapter examines the nature of governance in the specific context of regional planning in the Netherlands, which has had a Provincial administration system in place since its origin as a nation state in the sixteenth century, and now includes twelve Provinces. The Dutch formal provincial planning system looks neatly arranged between the national and municipal planning systems, having legal duties in many different areas, including physical planning, traffic and transportation planning, water management, and environmental planning. However, important questions remain as to how to deal with responsibilities that are shared with governmental authorities at other geographical levels. Similar questions exist with respect to relationships with non-governmental actors. Clearly, managing this complexity is not a simple and straightforward task (De Roo 2000).

The Dutch Provinces provide a variety of planning services, based on the four key areas indicated above. For example, they must translate national policies to the regional level and into guidelines for Municipalities; they must act to co-ordinate inter-municipal policy issues, for example in the field of housing; and they also have funds for implementing their own policies, such as the regulation of road traffic measures. However, their position is changing due to a more fundamental change in the orientation of Dutch planning from generic and centrally guided approaches towards area specific approaches, resulting in proactive development planning or project planning (see also WRR 1998, IPO 2001). The current Physical Planning Act is still very much geared towards centralised management and control. Experience in recent years has taught us that this approach is incapable of dealing with the many uncertainties involved in a more active project-oriented mode of planning. This background is prompting a series of transition at the regional level.[2]

The aim of this chapter is to explore some of the fundamental transitions taking place in Dutch regional planning, and to explain how these changes necessitate the increased involvement of stakeholders, actors and citizens. Firstly, the growing attention towards governance is considered (see also Voogd 2001b), and this is followed by a brief discussion (§ 4.3) about the role of regional government in so-called social dilemma situations: one of the more important aspects of a changing

---

2   For example, see Koeman (1999). Although the Netherlands has not experienced in the eighties the same 'Thatcherite revolution' as in the UK (including the disappearance of regional planning in some of the metropolitan areas) the debates around the renewed importance of regional planning show similar arguments (e.g. see Swyngedouw 1997).

and fuzzier regional planning picture. One of the key issues discussed is the need – due mainly to the complexity of the policy arena – to consult all of the parties with an interest in the plan and to identify the underlying motive or spirit behind policy statements, strategies and actions. The chapter provides an empirical illustration of the principles involved, based on an analysis of regional plans in the Province of North Holland (§ 4.4).

4.2 Regional planning: From government to governance

As indicated above, a primary aim of regional planning is to minimise any conflicting policies and practices across the local authorities within a region and to achieve a satisfactory relationship between people, jobs and the environment in support of a balanced regional development. The regional planning function is linked with a regional authority, be it a formal regional government, with elected members capable of making political decisions, or a coalition of local governments and/or other actors.

In the past regional planning was often primarily seen as an activity aimed at the production and review of some kind of 'regional plan'. Contemporary regional planning is much more than the application of traditional regional development theory (Cappelin and Batey 1993). It also includes social, political as well as organisational processes. For example, control of urban sprawl is impossible without an awareness of the social and geographical behaviour of the people living and working in the area, and without some form of regional political consensus about the need to contain it. This implies the need for regional co-ordination involving multiple local authorities and relevant actors. Arguments that underline this implication are easily identified. For example, environmental pollution does not respect local boundaries, and – likewise – other public affairs can often be best dealt with on a regional level such as, for instance, waste and water management. Globalisation processes also encourage a stronger regional representation in order to maintain regional distinctiveness on the European and worldwide stages. These and other phenomenon have the effect of encouraging a regional approach, which includes not only the governmental body itself, but also non-governmental actors.

The contemporary view of regional planning coincides with the proliferation of the word '*governance*' in recent years (see Chapter 3), which might best be summarised as the management of the common affairs of political communities by active collaboration of various interests. The most important characteristic of governance is that it involves more than government. It explicitly includes other actors who together attempt to determine the progress of developments in a region.[3] This effectively means a shift from a rather programmatic top down approach to policy, towards an approach that puts local and regional issues at the centre of the

---

3 From a theoretical point of view, governance is not a new idea. For instance, basic ideas can already be found in Paul Davidoff's (1965) famous article on 'Advocacy and Pluralism in Planning'.

stage along with all parties involved. The result of this shift is the rise in popularity of the area oriented or regional approach

Understanding regional governance involves both grasping the functioning of relevant regional and local institutions, the changing role of government and the operation of 'networks', both public and private, which attempt to control policy arenas. The emergence of the concept of *networks* is linked to a definition of regional governance as activities that emerge from the policy-making behaviour of political, administrative and societal actors on a regional level. The *network* paradigm has gained recognition in social, political and economic sciences as the theoretical basis for examining governance structures. A social network may include as 'nodes' a wide variety of actors: people, firms and other institutions. Between these 'nodes' different relationships can be distinguished. In network theory links, connections and interactions are seen as especially essential.[4] A regional social network becomes a regional *planning network* if it is used for selectively involving certain actors.

Governance implies co-operation of actors. Healey (1987) has developed a collaborative planning approach to the *design* of governance systems and practices, focusing on ways of fostering communicative, consensus-building practices. She addresses governance processes and the challenge of institutional design for collaborative planning. Healey distinguishes between two levels: the 'soft infrastructure' of practices for developing and maintaining particular strategies in specific places and the 'hard infrastructure' of the rules and resource of policy systems. The 'soft infrastructure' rests on communicative action theory (see also Innes 1995). Underpinning this theory is the assumption that 'reality' is socially constructed rather than being based upon 'facts' that can be known. Its epistemology is based on multiple forms of understanding that should be investigated through discourses. Woltjer (2000) has denoted this process as *consensus planning*. This may take multiple forms, in which communication, persuasion, learning, mediation, negotiation and bargaining are essential ingredients. In its application in planning practice, consensus planning can be used to deal with complex planning situations and to deliver public support, but it should not be regarded as a universally applicable instrument.

No doubt, effective communication is crucial for good governance. Communication involves two-way relations whereby both sides can share the facts and voice views, and whereby both sides can reframe their points of view to work towards a consensus. In this light, extra steps must be taken to facilitate the participation of actors, and to avoid a growing complexity and fuzziness. In terms of the 'hard infrastructure', governance involves creating new communicative platforms in which actors and citizens have an opportunity to participate in decision-making.

---

4   A rough distinction can be made between applications that focus on a quantitative, more formal, analysis of network structures (e.g. Wasserman and Faust 1994), and a more qualitative approach focusing on an understanding of the dynamics of policy-making processes (e.g. Teisman 1992, Klijn et al. 1995, Klein 1996). Especially the latter approach appears to be useful for understanding the mechanisms, possibilities and limitations of planning.

Dutch regional planning and decision-making are based on a model of representative democracy, which implies that the decision-making authority of the elected institution (e.g. provincial politicians) is of key importance. In a representative democracy, representation refers to structures of delegates who represent their people and use their skills to speak for people. Under governance, the emphasis on representation has shifted towards participation. Participation means, among other things, that people can speak for themselves (although in practice, most 'participation' tends to be through a selective and/or representative process). Decision-making processes thus no longer only refer to representing the absent, but now also refer to providing possibilities for direct participation (see Chapter 3).

The central premise seems to be that planning will be improved if actors and informed citizens participate in the process. One of the problems with the direct participation of active groups, however, is that policy formulation becomes too selective in terms of its participants and therefore restricts the depth of the debate (Akkerman, Hajer and Grin 2000). Another problem that comes with direct participation is who is going to participate and who is not. It is easy to say that all those parties that are somehow involved in matter should join in, however not all parties are equally willing to work towards consensus, and are willing to take its share of responsibility. Some parties might even obstruct planning processes. Selecting and inviting parties is a crucial process that has a tremendous effect on the project's success. It is also a fuzzy process, and is not easy to control.

Yet, it is obvious that the participation of a variety of stakeholders is essential to learn about community values and problems, and for avoiding unnecessary conflicts by making community members feel comfortable with the way the process is conducted.[5] It has been mentioned that planning networks should encourage civic engagement of actors. These networks help to facilitate co-ordination and communication, whilst – hopefully – encouraging a sense of trust between the actors. However, there is some evidence that networks will be less effective in larger communities, such as regions.[6] This is, of course, important for regional planning. It also reflects upon the need to identify appropriate tools for gaining information at the regional level, such as actor consulting (See Chapter 8).

---

5 The term 'community' is not unambiguous. Traditionally, the term refers to a set of people living in the same locality in which at least some resources are shared. Yet today, communities can range geographically from neighbourhoods to small rural hamlets to regional or even global scales. The term also suggests other definitions of an abstract nature in which physical boundaries are no longer implied (e.g. the 'Internet community'). In our mobile societies, it is possible to share resources with many different people in different places and thus be a member of many communities. 'Community', therefore, has become a loose concept; it is no longer a self-contained entity, which can be easily defined.

6 For instance, in his research on the decline of social capital, Putnam (1995) discovered that networks in larger communities are less balanced 'centres of discourse' than they are in small communities. This may be a weakness of regional governance. For a more elaborate discussion of communicative/ collaborative/ consensus planning: see Voogd and Woltjer (1999) and Woltjer (2000).

Notions and concepts such as networks and consensus planning play a key role in the current position of regional planning in the Netherlands. Influential studies such as those by the Dutch Scientific Council for Government Policy (WRR 1998), argue for a type of spatial planning that offers opportunities for the involvement of a large variety of actors – in particular at a regional level. According to the Council, national strategic planning policy should concentrate on main spatial structures and a restricted number of national projects. National government should limit itself to defining a set of so-called basic quality demands. The Provinces that in the past used to focus only on prohibitory planning, are now required to consider policies that support the physical development of regions within that province. At this regional level of scale – as the recommendations of the Council make clear – development plans should be formulated focussing on potential development opportunities taking in mind the specifics of the particular region to be addressed. The provincial administration implements these plans by means of allocating funding for key projects and by the designation of development areas. Evidently, by focussing on the development task of Provinces, the Council implicitly asserts that protective policies, such as those with respect to nature conservation, should be balanced against new developments.

As we will see below, a fundamental change that appears to be taking place in planning activities at the regional level is that the Provinces aspire to play a much more active and invitational role in development issues (IPO 2001). The implication of this is quite radical, in that the Provinces would gradually take responsibility for delivering the greater part of spatial planning in the Netherlands. The intention is that the Province steps out of its shell as 'reviewer and planner', and distinguishes itself as an 'entrepreneur' and a 'negotiator'. We will go on to explore some of the consequences of the changing role of actors and of actor participation as a result of a change in governance. This exploration indicates that these processes of change are far more fundamental to planning and to its fuzzy character than an expansion in the role of participating actors.

4.3   Social dilemmas: A support for government or governance?

A fundamental problem of consensus planning, regionalization, area specific approaches and hence of regional governance, is the existence of so-called *social dilemmas* (Voogd 1995, 2001a). A social dilemma is a conflict between the choice an actor would make to maximize his self-interest and the choice that would be best for the common good. In regional planning we are often confronted with situations in which private interests are at odds with public interests. A classical example is Hardin's 'Tragedy of the Commons'.[7] This is an example of a group of individuals who are faced with the problem of how to maintain their collective good and who

---

7   Hardin (1968) describes how individual farmers have access to grazing grounds that are held in common by all. Each can make a personal profit by adding livestock to the commons and each continues to do so. Adding an extra animal implies costs in terms of pasture consumed, but these costs are absorbed by the community. The tragedy comes about

– if they give in to their individual 'greed' – produce a collectively undesirable outcome.

In order to prevent such a situation, individuals have to restrain themselves. This is not an easy affair. Any individual in the above situation may consider two possibilities. First, if others do exercise restraint, one can personally enjoy the fruits of their restraint without having to contribute to its costs. That is, by being a so-called 'free-rider', one profits from the fact that others may be prepared to take action to prevent an undesirable situation. Alternatively, one may consider the possibility of being a 'sucker' who incurs a cost when nobody else does.

The implications of this behaviour can have far reaching effects for all aspects of life. Forrester (1969) researched this phenomenon in the 1960s. His study of 'system dynamics' led him to believe that some of the causes of pressing public issues are grounded in the very politics intended to solve them. These policies are too often the result of superficiality and social acceptability, which devise interventions that focus on obvious symptoms, instead of the underlying causes. This produces short-term benefits, with little or no long-term improvement.

Different types of social dilemma can be distinguished, for instance based on the number of people in the group, the nature of *contributions* (continuous or once-only), the nature of rewards or utility, and so on (see Axelrod 1984). An interesting aspect of social dilemmas from a planning point of view relates to the quality of the initial collective situation, typified respectively by the *provision* and the *maintenance* of a public good. The social dilemma of Hardin's metaphor refers to a dilemma in which the initial situation is desirable from a collective point of view. But it is dependent upon individual actions to keep it desirable, in other words, to maintain the current situation. Examples are all around us: the protection of our cultural heritage, nature areas, historic town centres, and so forth. But we can also distinguish situations where the initial situation is an undesirable one: derelict land, heavily polluted areas, areas with urban blight, and so on. This demands costly individual action to transform the situation into a desirable one from a collective point of view.

Therefore, in social-dilemma situations, representative government may sometimes have to intervene in favour of the collective public interest. Theory and practice feature very different expressions of the term 'public interest' (Gilbert 1979, Berkowitz 1979). However, some general conditions must prevail if any public policy can be said to be in the public interest. These conditions include a need to maintain a balance of different interests. The collective interest lies in the quality of the environment for current and future generations and stretches beyond the satisfaction of individual needs. In a decision-making process, people may consider the collective interest but, when actually asked to contribute, consider self-interest. For instance, local governance may be at odds with interests at a higher (regional) level of government. Moreover, a common problem related to governance is that future generations can never directly participate. Governance processes may

---

because every farmer is inclined to increase his own herd, while leaving the costs of grazing to be borne collectively. Ultimately the commons will be destroyed.

therefore tend to exclude their interests. Clearly, in social dilemma situations weak interests require special protection, and collective interests cannot be protected by a governance approach alone.

Social dilemmas can only be solved by some kind of mutual restraint.[8] Research of social psychologists into this matter did show that in general a stakeholder has a strong reluctance to hand over his freedom of decision to a higher authority (see for example Wilke et al. 1986, Rutte et al. 1987). This reluctance can be overcome, however, when a stakeholder is sufficiently *compensated*. A certain loss of control may be acceptable when a stakeholder is faced with a situation in which there is a threat of a collective catastrophe or when a social disadvantage may be threatened. Under these circumstances, decision-making by a higher authority may become acceptable.

Given these research results, it would be a logical next step to proclaim regional government as a higher authority that should solve social dilemmas at the local level.[9] For example, high-quality public transport would never reach the thinly populated periphery of a country without national and regional government support (see Vlekken and Keren 1992). These observations support the conclusion that the importance of the role of a regional *government*, in its capacity for solving social dilemmas, can never be neglected in regional governance.

An important step to overcome these dilemmas is to achieve a better understanding of interests, motives, perceptions and behaviour of the various actors involved. Actor-consulting is a means to visualise these aspects and as such can sharpen the institutional context in which decisions are made.

4.4 Regional plans in the Province of North-Holland

The changing complexity and the rise of fuzziness or fluidity of regional planning in the Netherlands can be illustrated by considering a series of three regional plans in the Province of North-Holland. In the last decade of the twentieth century, Dutch regional plans have shifted from a strict emphasis on prohibitory planning towards more emphasis on a development-oriented type of planning (see Woltjer and Meijnen 2001). This shift towards development goes hand in hand with an increasing emphasis on regional governance. Regional planning in North-Holland features an increasingly active, governance-oriented provincial government. The Province is gradually giving more attention to establishing structures to facilitate collaboration and to developing networks of actors in relation to specific projects.

---

8 According to Hardin (1968) solutions that do appeal to individual cooperation are doomed to be unsuccessful.

9 However, as elaborated in Voogd (2001b), this is less obvious than it may seem. It depends on the quality of government whether they are willing and able to recognise social dilemmas and whether they want to act accordingly. This recognition of social dilemmas by the government is very important, especially if we accept the results of psychological research that in case of social dilemmas voluntary cooperation alone does not work.

The *Waterland regional plan* was adopted in September 1990. The key planning issues included the high demand for housing due to shortages in Amsterdam, the bleak expectations for the agricultural sector, threats towards landscape and nature, and shortages in recreational facilities. The general objective for Waterland is creating conditions under which the rural area can subsist in its current form and function. The policy ambition can be summarised as guarding rural areas from urban functions such as housing and industrial estates, and at the same time concentrating these functions in existing urban areas.

Figure 4.1   North-Holland South plan area

The *Kennemerland regional plan* was approved in December 1998, and was drawn up to set a coherent policy for the activities around the Amsterdam-North Sea Waterway, the channel linking the harbour of Amsterdam with the North Sea. Key planning issues here included the need to provide a coherent plan to link areas of ecological value, the need to elaborate on house-building projects prescribed by national government, the increasing demand for industrial parks, and the accessibility of the coastal areas. The overall objective for Kennemerland is to work towards vigorous/ vital residential areas and work areas in a green environment. The core of the policy aims to create a balance in living, working and nature by means of zoning.

Overall, we can say that we are dealing with an 'old style' plan for Waterland, whereas the 'new style' plan for Kennemerland was realised in a more collaborative way. That means – among other things – that citizens and other actors were involved more intensively in the plan-making process of Kennemerland. Even though the participants in the plan-making processes of both plans are virtually the same, with Kennemerland they were dealt with more directly and interactively. Kennemerland distinguishes itself through an intense involvement of citizens, and the description of some of the topics makes clear that intensive rounds of discussion took place between the Province and the main stakeholders, including for example, environmental groups and the chambers of commerce. In contrast to this approach, Waterland restricted itself to the involvement of the provincial and municipal government agencies only, while legally prescribed consultation activities played a central role.

The outcome of the planning process has been particularly positive in the case of the Kennemerland plan, where, not only the Province, but also the target groups of the consultation exercise, such as Municipalities, interest groups, and citizens have freely deployed the plan in their own activities. The fact that various actors feel that they are 'owners' of the plan can be traced back to the collaborative plan-making process.

In addition to these governance-based characteristics, regional plans can also be considered in terms of their *functions*. A regional plan will generally include the *development function*, the *integration function*, and the *conformance testing function*. The development function gives emphasis to the regional plan as a stimulator for physical and economic developments, initiatives, projects, and ideas. The integration function underlines the purpose of a regional plan to unite various sectoral policy fields (horizontal co-ordination) and spatial planning policies at different government levels (vertical co-ordination). With regard to the conformance testing function, regional plans provide a vehicle for Provincial government to evaluate lower level spatial plans, to check if all actions taken are in line with plan policies and proposals. Although historically the main function of Dutch regional plans has been to provide a conformance testing framework for the control of municipal land use plans, an important question remains as to whether the activities of the target groups (municipalities, interest groups, citizens) fit within the scope set out by the regional plan.

Let us now look at the functions of the Waterland regional plan. The Waterland plan generally provides an indicative sketch of desired spatial arrangements in the region. The main function of the plan is to serve as a basis for conformance testing the municipal land use plans, and consequently it does not devote much attention to development and implementation characteristics. For example, the Waterland plan indicates development projects for housing near the city of Purmerend, but the final housing layouts are dependent on further collaboration with the municipality, real estate developers, and interest groups. However, there is some concern that the conformance testing activities by the Province for the Waterland region have left little room for creativity in municipal plans. Similarly, for Kennemerland, reviewing of municipal plans against the regional plan by the Province has resulted in the development zones being strictly implemented. The Municipalities are consequently concerned that their locally oriented policies are being undermined by the strategic approach of the regional plan.

In summary, the important differences between the plans are related to function and to governance characteristics. In particular, the extent of collaboration involved in the development of the Kennemerland plan is associated with more development-oriented lines of policy, and the actors feel a greater ownership of their regional plan.

The *North-Holland-South area* is a part of the northern wing of the urban 'Randstad', the urban zone covering Rotterdam, The Hague, Amsterdam, and Utrecht. The final version of the 'North-Holland-South' regional plan was adopted at the end of 2002. The plan includes the Waterland and Kennemerland regions, and will therefore serve as a replacement of these plans. The central objective is aimed at preserving and strengthening the northern part of the Randstad as an economic driving force (with the exception of Waterland), while paying attention to demands for a good quality of life, accessibility, and water management. The goal is to create more coherence in the spatial policies that were scattered over eight different regional plans.

The new plan features a collaborative, strongly governance-oriented plan-making process (PNH 2001). A lot of attention is paid, for instance, towards opportunities for citizens, municipalities, private companies, and institutions to take part in dialogues about the interpretation of the plan. The Province aims at taking in the ideas of citizens and stakeholders to develop public support. The wishes and desires of these actors become the centre of attention, while less emphasis is put on offering policy choices in a 'top-down' fashion. The outcome of the process is expressed in the form of a 'citizens recommendation' for the new plan. Collaboration may be facilitated by discussion evenings, internet sessions, or 'kitchen table conversations' (informal discussions).

The provincial government remains in control of strategic policy choices for developments at the supra-local (regional) level of scale. Furthermore, the Province sets certain limits for the extent to which the voices of municipalities and other stakeholders can be incorporated into the plan. Clearly, municipal government remains the authority that can solve social dilemmas at the local level. However the provincial government has to make decisions with regard to issues which may sometimes be in conflict with local needs, including for example, setting coherent ecological structures, clustering industrial zones, or protecting open landscapes from ad-hoc housing plans.

Another important principle for the new regional plan is that it has to be on the one hand broad, yet on the other hand it has to provide a clear framework within which the parties involved can develop policy. The central idea is that the North-Holland-South plan provides a policy outline, and that the details are omitted, based on the argument that adequate space should be left for custom-made initiatives at the Provincial and municipal level. Consequently, room is made for initiatives to be brought in by the various stakeholders, on the basis of mutual trust and consensus.

4.5   The effective delivery of quality in planning

So what do these reflections upon regional planning imply? It is clear that in the course of the 1990s, the functions of the regional plans in the Netherlands have changed. This change is part of a broader shift in planning theory and practice. There is a growing priority for the development function of the provincial plan, a diminishing importance of the conformance testing function, and a re-orientation of the integration function. It leads to area-specific approaches and to the interaction and involvement of other actors. Our series of three regional plans in the North-Holland Province illustrate this shift. The character of regional planning has changed towards more dynamics, more flexibility, more collaboration, and – predominantly – towards a more development-oriented planning style. The involvement of actors outside the governmental domain has become an essential part of the planning process. Regional government involvement not only refers to prohibiting the undesirable, but increasingly to stimulating desirable developments of a high quality. The idea is that 'good quality' can only be attained when regional government is engaged in close collaborative relationships with other actors – be it public or private parties.

Overall, the foundation has been laid for a type of development-oriented governance that strongly links plan-making and implementation. As touched upon, the notion of 'spatial quality' plays an important role. In regional governance, there should be room for situation-specific considerations aimed at quality improvements. The notion of spatial quality relates to creating coherence between providing space for social activities, and the perceived values that result. The recent proposals for changes of the Dutch planning law (VROM 2001), for instance, indicate as a central consideration that new rules are given for the advancement of spatial quality and that planning should ensure a 'sustainable' environment.

Municipalities are usually able to quote examples in which a local planning proposal could potentially deliver an improvement in spatial quality, but the proposal is not viable because of a mismatch with the regional plan (Woltjer and Meijnen 2001). These examples usually deal with small projects such as the displacement of companies in the countryside, or the joint use of nature and recreational activities. Accordingly, it is argued that decision-making should be more flexible. This means that a local plan should not be tested on a stringent 'yes' or 'no' basis, but rather upon the underlying essence of the regional plan. In Chapter 2 Healey points out that a restructuring of the frames of reference by the actors is desirable in these circumstances. The question is not only important, therefore, as to whether the objectives from the regional plan are used for testing the conformance of local policy. Equally important is whether these objectives are actually desirable in a local context, or whether they are understood and succeed in matching the frames of reference of other crucial actors. We can denote this way of thinking under the heading 'substantive rationality' (Mannheim 1940). Many parties that were affected by or involved in the Waterland and Kennemerland regional plans have frequently called the objectives into question. The regular occurrence of substantive rationality critique on some of the goals of these regional plans underlines the importance of argumentation.

This would mean, among other things, that the regional government has to play a role in either directly consulting relevant actors, or facilitating a discussion of the diversity of problems and interests that are involved. For the Waterland and Kennemerland regional plans, it was clear that parties could not detect adequately the argumentation underlying the various objectives (Woltjer and Meijnen 2001). This is important if one wants to develop a type of planning that matches a dynamic society and delivers geographically relevant qualities. Therefore, the objectives that are specified in the regional plan should not always be taken for granted. The underlying argumentation as to why the objectives should be defined in this way must be open to both consultation and debate on an ongoing basis, with the Province or regional authority in a co-ordinating and mediating role.

4.6 The role of the controlling authority

It has been argued in this chapter that certain social dilemmas call for 'top-down' government. This problem is usually dealt with by introducing the *subsidiarity principle*, whereby the decision-making authority should reside at the level most appropriate to the problem being addressed.[10] Unfortunately, this is not an unambiguous criterion, as discussions regarding the European Union clearly illustrate (for example, see Toulemonde 1996).

Three elements can be distinguished that characterise the role of the controlling authority. The first element concerns the level at which a planning authority should be vested with powers for developing certain policy areas. This relates to the well-known debate over *centralization and decentralization* (see for an overview Eigenberger 1994 and De Roo and Voogd 2004). Decentralization can allow the government to respond more effectively to variations in local needs and preferences. This may in turn provide opportunities and incentives for policy innovation; and give citizens greater choice and voice in policymaking. Centralization, on the other hand, enables government to address problems characterised by cross-border effects; exploit available economies of scale; co-ordinate policies more effectively; and promote equality and political homogeneity across a larger domain to reflect 'shared values'. The respective advantages of centralization and decentralization therefore suggest that the optimal allocation of authority may prove to be highly context-dependent. The second element regards the question of how intrusive the policy-making of a regional authority should be. Clearly, it should not render local policy-making superfluous and the demands and needs of local actors must be addressed. The third element addresses the question of how individual stakeholders can retain some control over a policy area that is delegated to the regional authority. This calls for dialogues and consultations in a network context.

A clear thread in the Waterland, Kennemerland and North-Holland-South regional plans is the increasing attention to the development function of regional plans. A development-oriented regional plan typically restricts itself to sketching some main

---

10 In the Treaty on the European Union, Article 3b, it is described as: "In areas which are not under its exclusive power, the Community shall act in conformity with the principle of subsidiarity, only if and in so far as the objectives of the proposed action cannot be sufficiently achieved by the member states and can therefore, by reason of the scale and efforts of the proposed action be better achieved by the Community".

outlines only, whilst providing a means to bind, stimulate, and motivate various actors. It leaves space for the development of all kinds of innovation. An increased focus on stimulating development would mean that the testing of conformity of local spatial plans against the regional plan takes place more on the basis of arguments about spatial quality. This implies using the intentions, and not only the specific rules, of the plan as a touchstone. If a local proposal is to be tested against the regional plan, then the motivation and the interpretation applies rather than the exact content of the proposal. In some cases, however, there is a case for a strict review of local plans, because 'weak' interests that need protection, such as certain environmental issues, may indeed require a stringent assessment.

The integration function of regional plans facilitates the joining together of a variety of spatial policies into one coherent plan. Regarding the North-Holland situation, this type of integration seems to have become a matter of co-ordinating spatial planning, environment, water management, and transportation for specific regional or local projects. Integration takes place in an area-oriented way or by means of a specific development project.

Our reflections suggest that a development-oriented regional plan should contain some criteria or standards about *spatial quality* against which local proposals can be judged. These criteria then provide a touchstone for setting up consultations and discussions aimed at the development of projects or land, in which actor consulting might play an important role. This means that substantive intentions can be of more importance than the means to facilitate a clear-cut decision. These remarks call for an explicit warning, however. If a surplus of attention were devoted to development processes, all spatial planning could end up being determined by active participants only, and a coherent strategic policy could become a thing of the past. So it might be true to say that the protection of collective interests at times calls for controlling actions by regional government.

Furthermore our argument indicates that broad quality criteria should not be in force for a whole regional planning area, and that a made-to-measure local approach is crucial. This might involve, for example, a type of regional planning that distinguishes between areas in which orders and prohibitions are strictly indicated, and areas in which freedom is provided for actors to find optimum solutions related to all interests involved. The last category would need to be based on criteria that make clear what the quality demands are for allowing development, and how certain land uses can be combined.

The weight of evidence therefore suggests that regional planning should always offer the opportunity for a consensus planning process involving a wide range of actors. Our argumentation however supports the view that a governance approach to regional planning that partly preserves some 'government' is most likely to deliver effective regional development. It has also been reasoned that regional governance should be regarded as a way of decision-making that should enrich, not replace, existing democratic and representative institutions of government. In this environment, techniques that help to develop an understanding of the various frames of reference of actors are consequently likely to be of significant value.

The complexity of regional planning inevitably leads to active involvement, with consequent opportunities for actor consulting.

References

Akkerman, T., M. Hajer and J. Grin (2000) 'Interactive Policy Making as Deliberative Democracy? Learning from New Policy-Making Practices in Amsterdam', paper presented at the Convention of the American Political Science Association, Washington DC.

Axelrod, R. (1984) *The evolution of cooperation*, Basic Books, New York.

Balchin, P., L. Sýkora and G. Bull (1999) *Regional Policy and Planning in Europe*, Routledge, London.

Berkowitz, I. (1979) 'Social Choice and Policy Formulation: Problems and Considerations in the Construction of the Public Interest', *Journal of Sociology and Social Welfare*, Vol. 6(4), pp. 533–545.

Buiten, J., W. Huizing, F. Osté, G. de Roo and H. Voogd (2001) 'Regional Planning and Sustainable Development: Introducing an actor-consulting model', Dutch case study of the Intereg IIC Susplan project, Physical Planning and Environment Department, Province of Drenthe and Faculty of Spatial Sciences, University of Groningen, Assen / Groningen (NL).

Cappelin, R. and P. Batey (eds) (1993) *Regional Networks, Border Regions and European Integration*, Pion, London.

De Roo, G. (2000) 'Environmental conflicts in compact cities: complexity, decision-making, and policy approaches', *Environment and Planning B: Planning and Design*, Vol. 27, pp 151–162.

De Roo, G. and H. Voogd (2004) *Methodologie van planning – over processen ter beïnvloeding van de fysieke leefomgeving* [Methodology of planning – on processes influencing the physical environment], Uitgeverij Coutinho, Bussum (NL).

Eichenberger, R. (1994) 'The Benefits of Federalism and the Risk of Over-centralization', *Kyklos* 47, pp. 403–420.

Forrester, J.W. (1969) *Urban dynamics*, M.I.T. Press, Cambridge (US).

Gilbert, N. (1979) 'The Design of Community Planning Structures', *Social Service Review*, Vol. 53(4), pp. 644–654.

Hardin, G.R. (1968) 'The tragedy of the commons', *Science*, Vol. 162, pp. 1243–1248.

Healey, P. (1997) *Collaborative Planning, shaping places in fragmented societies*, Macmillan Press, London.

Healey, P., A. Khakee, A. Motte and B. Needham (1997) *Making Strategic Spatial Planning: Innovation in Europe*, U.C.L. Press, London.

Innes, J.E. (1995) 'Planning Theory's Emerging Paradigm: Communicative Action and Interactive Practice', *Journal of Planning Education and Research*, Vol. 14(3), pp. 183–89.

IPO (2001) 'From ordering to developing, Provinces invest in spatial quality', Advice of the IPO Spatial Development Politics committee, Inter-provincial Association, The Hague (NL).

Klijn, E.H. (1996) 'Analysing and Managing Policy Processes in Complex Networks: A Theoretical Examination of the Concept Policy Network and its Problems', *Administration and Society*, Vol. 28(1), pp. 90–119.

Klijn, E.H., J. Koppenjan and K. Termeer (1995) 'Managing networks in the public sector: A theoretical study of management strategies in policy networks', *Public Administration*, Vol. 73(3), pp. 437–454.

Koeman, N.S.J. (1999) 'Expectation: a fundamental revision of the Physical Planning Act, *Nederlands tijdschrift voor bestuursrecht*, Vol. 4, p. 89.

Lefèvre, C. (1998) 'The idea of governance: an attempt at clarification, Transport and land-use policies: resistance and hopes for coordination', proceedings of the launching seminar of the Action COST 332 24–25 October 1996, Barcelona, European Commission, Directorate General Transport, pp. 63–72.

Mannheim, K. (1940) *Man and society in an age of reconstruction*, Routledge & Kegan Paul, London.

Martins, M.R. (1986) *An Organisational Approach to Regional Planning*, Gower, Aldershot (UK).

Putnam, R.D. (1995) 'Bowling alone: America's decline in social capital', *Journal of Democracy*, Vol. 6(10), pp. 65–78.

Rutte, C.G., H.A.M. Wilke and D.M. Messick (1987) 'The effects of framing social dilemmas as give some or take some games', *British Journal of Social Psychology*, 26, pp. 103–108.

Self, P. (1980) 'Whatever Happened to Regional Planning?' *Town and Country Planning*, Vol. 49(4), p. 209.

Swyngedouw, E. (1997) 'Neither global nor local: "globalisation" and the politics of scale', in: K. Cox (ed.), *Spaces of globalisation: reasserting the power of the local*, Guildford, New York, pp. 138–166.

Teisman, G.R. (1992) *Complex decision-making, a pluricentric perspective on decision-making about spatial investments*, VUGA, The Hague (NL).

Toulemonde, J. (1996) 'Can we evaluate subsidiarity? Elements of answers from the European practice, International review of administrative sciences', Vol. 62, pp. 49–62.

Vlek, C.A.J. and G.B. Keren (1992) 'Behavioral decision theory and environmental risk management: assessment and resolution of four 'survival' dilemmas', *Acta Psychologica*, Vol. 80, pp. 249–278.

Voogd, H. (1995) 'Environmental Management of Social Dilemmas', *European Spatial Research and Policy*, Vol. 2(1), pp. 5–16.

Voogd, H. (2001a) 'Social Dilemmas and the Communicative Planning Paradox', *Town Planning Review*, Vol. 72(1), pp. 77–95.

Voogd, H. (2001b) 'Rethinking Regional Governance and Planning', in: Zoltan Gal (ed.), *Role of Regions in the Enlarging European Union*, Centre for Regional Studies of the Hungarian Academy of Sciences, Pecs, Hungary, pp. 194–213.

Voogd, H. and J. Woltjer (1999) 'The communicative ideology in spatial planning: some critical reflections based on the Dutch experience', *Environment and Planning B: Planning and Design*, Vol. 26(6), pp. 835 – 854.

Wasserman, S. and K. Faust (1994) *Social Network Analysis: Methods and Applications*, Cambridge University Press, Cambridge/New York.

Wilke, H.A.M., W. Liebrand and D.M. Messick (1983) 'Sociale Dilemma's, een overzicht' [Social Dilemmas, an overview], *Nederlands Tijdschrift voor de Psychologie*, Vol. 38, pp. 463–80.

Wissink, B. (2000) 'Provinciale omgevingsplannen ter discussie, kanttekeningen bij een oude gedachte' [Provincial comprehensive plans under discussion, comments on an old concept], ROM Magazine, Vol. 12, pp. 13–16.

Woltjer, J. (2000) *Consensus Planning, the relevance of communicative planning theory in Dutch infrastructure development*, Ashgate Publishing, Aldershot (UK).

Woltjer, J. and R. Meynen (2000) 'Evaluatie streekplannen Waterland en Kennemerland' [Evaluation of the Waterland and Kennemerland Regional Plans], report of a study funded by the Policy Evaluation Committee of the Province of North Holland, Haarlem (NL).

VROM (2001) 'Voorontwerp voor een Wet ruimtelijke ordening' [Preliminary Design for a Physical Planning Act], Explanatory Memorandum, August 2001, Ministry of Housing, Spatial Planning and Environment, The Hague (NL).

WRR (1998) 'Ruimtelijke Ontwikkelingspolitiek' [Spatial Development Policy], *Netherlands Scientific Council for Government Policy*, 53, Sdu Publishers, The Hague (NL).

Chapter 5

Indicators in the Context of Fuzzy Planning

Donald Miller[1]

5.1 Introduction

Spatial planning requires the insights and contributions of all affected parties – the 'actors' discussed in this book. This is not only an important requirement of democratic government but, as already pointed out, it is required because each of these parties has perspectives and unique information which are needed to identify a full set of goals and means, and because strong endorsement of these by affected parties is the basis of the political support needed to implement plans once they are approved (McDonald 1996).

A key technique to facilitate this process of engaging the actors who are central to success in planning is to develop a set of measurements that represent the goals and objectives sought, and which can be used to gauge how well the means might serve to accomplish the goals. These measurements are commonly known as indicators. They differ from raw data in that they are selected as instruments for thinking about and representing the purposes which planning is intended to serve, for evaluating alternatives in terms of their effectiveness in achieving these purposes, and later in monitoring and evaluating how successful the selected alternative has proved to be. In short, it is difficult to conceive of a purposeful and thoughtful planning and implementation process that does not involve the design and use of indicators in some capacity. While indicators do not have the power to guarantee full objectivity often assumed by many, these might add to a better understanding of objects, targets and progress in planning, and might diminish planning's illusive or fuzzy character in complicated and complex situations.

Sustainability is considered in this book to be an example of one of planning's most complex, underestimated and misunderstood notions. It is used in this chapter to provide a context for information that is or should be made available for decision-making purposes in the planning process. Sustainability is seen as a *fuzzy* notion, being illusive in its meaning and its operational value. To make information about sustainability manageable the focus will be on indicators, and how these indicators relate to the actors involved.

---

1 Donald Miller is Professor in Planning at the Department of Urban Design and Planning of the University of Washington, Seattle, US.

Following a more formal definition of sustainability indicators, this chapter considers their uses, the process of their development, and the role of the participants involved. The chapter concludes by suggesting criteria for designing and using a set of sustainability indicators, seen from both a *pragmatic* and *communicative* perspective. In particular the *pragmatic* perspective relates indicators and their informative role to the actor-consulting model, which is at the core of this book.

5.2   Sustainability indicators: The traditional perspective

Indicators are measures of qualities or quantities that are useful in gauging the condition of something of interest. Sustainability indicators focus on the elements of what we wish to conserve and enhance in the setting of a functioning society (Hart 1995). From a functional, and therefore rather traditional perspective citizens must have some understanding about what these indicators should represent.

As Beatley (1995) argues, sustainability emphasizes the decisions and actions that contribute to the long-term viability of human activity. For most commentators, this means not only concern for natural environmental features, but also for the economy and features of society that are important over time (World Commission 1987, Berke and Conroy 2000). Environmental features include important resources that provide inputs for production, disposal of wastes, and the health of ecological systems. The intention is to assure that a set of physical stocks and conditions are available to future populations, which will accommodate the potential for economic and evolutionary development comparable to current levels (Lafferty and Meadowcroft 2000).

Economic features of sustainability include the wise use of non-renewable as well as renewable resources, in a manner that supports continued employment and income, balanced with the other needs of society and the environment. The social dimension of sustainability addresses the institutional arrangements contributing to the well-being of residents of an area, and the fairness with which opportunities and resources are distributed. Planning for sustainable development involves balancing environmental protection, economic development and social arrangements, rather than allowing them to come into conflict with each other (Kaiser et al. 1995, Campbell 1996, Beatley and Manning 1997).

From this traditional perspective attempts are made to give sustainability indicators sense and meaning. This is illustrated by the three perspectives that are common in viewing land and its development: exchange value, use value, and environmental value. *Exchange value* is the most conventional of these, and has to do with how land can be developed to produce the greatest net income (Logan and Molotch 1987). Thus it is the market value in monetary terms, and tends to cause owners to exercise a short-term horizon in their decision-making (Nelson 1986).

In contrast, *use value* of land refers to its social utility in accommodating human activities of all sorts. While exchange value tends to emphasize the private-good or commodity features of property, social purposes to which land may be allocated include common-good or communal features of land, such as cultural features and places for gathering and for joint activity.

Finally, *environmental value* of land is concerned with its role in supporting natural systems, providing habitats and even providing environmental services (Freeman 1993).

Given the fact that we are reflecting here upon an attempt to give indicators meaning from a traditional, technical and functional perspective, in the sense that it expresses a belief that sustainability can indeed be valued objectively, puts the three perspectives expressed here into a different and critical light. When taking a closer look it is not hard to identify the differences in view, behind these perspectives, which express various interests and truths. Thus, the exchange-value perspective represents a materialistic view of land as a commodity, the use-value perspective perhaps represents an anthropocentric view, and the environmental-value perspective may be eco-centric. From the functional or technical perspective, the task of planning for sustainability is to balance these three views of land in a manner that avoids long-term, consequential and irreversible outcomes such as loss of species or harm to humans (Corson 1995, Gouldson and Murphy 1997).

As will be seen in Chapters 6 and 7, this somewhat technical approach might under straightforward planning conditions be entirely appropriate, as long as its implications are recognised. However, the world around us is often far from objective. This means, we have to recognise that perceptions, belief systems and uncertainty about the future are also part of reality. Also for 'sustainability' values, it is not difficult to identify the underlying perspectives and interests – referred to as 'frames of reference' in previous chapters – from which these values are viewed. In the sections to come, various options will be discussed to overcome the technical and 'objectivity' trap into which sustainability indicators might lead us.

5.3 Purposes of sustainability indicators

As discussed at some length in Chapter 1, sustainability is illusive in terms of practice (see also Haughton and Hunter 1994). A major use of sustainability indicators is to translate abstract statements of goals or of a vision of a desired, long-term future into concrete, operational terms. This is the approach taken by Viborg Province, as described in the case study of Chapter 9.

These goal-oriented measurements can play important roles in designing and evaluating courses of action in a plan-development process, in monitoring and reporting on the effectiveness of these courses of action once they are implemented, and more simply in reporting on the changes taking place over time in the built and natural environments.

Sustainability indicators can be used to provide *guidance* in preparing plans and designing programmes. If these indicators are translated into equivalent goals or objectives, they can provide focus in designing planning options and in turn serve as criteria in evaluating these options to assist parties to the decision process in making choices (Common 1995). In this guise they can provide a tangible agenda for the process of developing plans, and can provide the information base for public deliberation aimed at reaching widely supported policy agenda (Reardon and Shields 1997). A potential drawback here, however, is that attention may be focused upon the indicator itself, instead of the underlying issue that the indicator is seeking to represent.

A second purpose of sustainability indicators is for use in *monitoring the performance of plans and programmes* (Deyle and Smith 1998). While private firms are required to prepare and release performance information on a periodic basis, similar requirements for evaluation by governmental agencies and other public institutions are increasingly widespread (Troy 1996, Howe 1993). Growth management legislation in the U.S., for example, commonly provides that cities and counties must develop a monitoring system for tracking development as it takes place, and for assessing the effectiveness of programmes undertaken to implement long range plans (Nelson and Duncan 1995). Measured progress toward goals is both useful in adjusting plans and programmes as conditions change, and in providing accountability to the public in terms of the effectiveness of these policies and actions (Freilich 1999, Burby and May 1998).

A third purpose of sustainability indicators is to provide periodic reports to stakeholders concerning changes that are taking place, whether these changes are improvements over prior conditions, and the rate of progress toward targets. Development and use of measurements in this manner may be undertaken either by government agencies, or by other non-governmental actors (Yap 1989).

A fourth purpose of sustainability indicators, related to the first three, is to provide a basis for a development management strategy (Margerum and Born 1995). In this application, these indicators are each related to policies and are used as performance measurements in selecting and adjusting actions to implement these policies (Smith 1998). Since these indicators are used as specifications concerning desired changes, this approach does not presuppose the design of programmes or projects, as might be the case with a formal plan, but rather encourages creativity in exploring ways of most efficiently responding to these policies. In this manner, the indicators provide a framework for an incremental approach to successively implement adopted policies, or make desired progress toward targets, modifying these actions when new evidence indicates performance that is unacceptable. As a management strategy, this approach is similar to performance contracting. It is illustrated by performance regulations for land development, in that courses of action are not prescribed but rather incentives and requirements are negotiated within a set of parameters. A management strategy of this sort promotes feedback and adjustment, which builds on learning and experience and helps to address uncertainty, fuzziness and illusiveness (May et al. 1996).

A final issue concerning the application of sustainability indicators is whether or not to specify targets for accomplishment. Establishing targets such as specific improvements in air quality or the numbers of units of affordable housing by a stated date heightens the sense of purpose and represents resolve which can animate a planning process and focus public interest on ways to accomplish these ends.

On the other hand, targets as commitments can limit the flexibility of elected decision makers to deal with other demands for resources that unexpectedly arise after the targets are set (National Research Council 1999). At the least, it is important to assess the costs of means to reach each target, and indeed to assess the costs of dealing with interactions between the features of sustainability that are being addressed at the same time, since some of these goals may not be mutually supportive (Davies 1987). These relationships between various goals may be easy to recognize, while remaining difficult to estimate accurately in terms of costs (Abramovitz 1997). Once costs are estimated, questions concerning funding are posed. These may lead prematurely to rejecting some targets, or rejecting programmes under consideration for making progress toward these targets (Bartelmus 1999). Finally, targets tend to be viewed as minimum levels of acceptable change or performance. Often, however, these floors for assessing accomplishment in practice become ceilings since progress beyond them is seen as less important. This is a problem shared with the use of standards in many situations.

Some of these difficulties can be avoided if targets are seen as representing the direction of progress rather than as specifications of success. It may be difficult, though, to convince the interested parties that targets are primarily an aid to direction-setting, and that progress short of hitting them is what is sought. Even if targets are not established, other reference points can be used to assess whether progress is being made and its rate. Examples include movement away from conditions existing earlier in a community, or in comparison with situations in other communities. Achieving movement towards a goal over time avoids the inflexibility of specifying particular levels of accomplishment at particular future dates.

5.4 Processes for developing sustainability indicators

Sustainability indicators designed for the four types of applications just discussed may differ concerning whether their technical sophistication and validity is more, or less, important than public understanding and acceptance (Pitts 1997, Miller 1997). Technical quality may, for example, be especially important for indicators used in a management strategy. In this context, some of the indicators may be engineering specifications or environmental features such as parts of various pollutants per volume in air or water quality measurements. In this situation, public involvement may centre on identifying the concerns to be addressed, with the consequent design of the measurements being largely scientifically based and the responsibility delegated to experts (Limoges 1993). The case study at Chapter 9 follows this pattern.

In the more common case, these responsibilities are shared between scientifically trained professionals and the public (Lynn and Kartez 1995). Actors may demand to be directly involved so that they understand the development and use of these measurements, and to confirm that their concerns are being addressed (Daniels and Walker 1996). Experts are important resources in this process, but often need the assistance of planners as brokers of this information between technical and lay participants.

Politically, citizen groups not only frequently insist on being directly involved in this process, but also by being involved they begin to take ownership for their own contributions and to develop a sense of advocacy for the measures which result. When this happens, these participants become a constituency for the indicators programme and gain confidence in employing them in deliberations leading up to decisions (Environment Agency 1998). As a result of this dynamic, coalitions can form which support a factually-based process of decision making, and which consequently support plans and programmes that address the issues represented by the indicators (Lynn and Busenberg 1995).

This is illustrated by the example of Sustainable Seattle (1993). The process used in this case involved informal workshops, a civic panel, and public forums over a period of three years. At one point, ten teams organized by topic had drafted 99 indicators. By the seventh draft iteration, this list had been consolidated to the 40 indicators used in the series of reports published by this group on the state of sustainability in the region (see Table 5.1). While scientific validity was a major criterion used in enumerating and specifying indicators by participants in the Sustainable Seattle programme, the meaningfulness of these indicators to non-technical parties was of even greater importance. Consequently, a major item of discussion throughout this process included choosing the most important features of sustainability for the final list, and choosing measures for these that residents could relate to on the basis of their own experiences.

For instance, the size of runs of wild salmon returning to spawn in regional rivers and streams is used as a measure of water quality and the state of riparian habitat. Water quality experts regard this as a crude index since it does not distinguish between pollutants that may be present, and marine scientists point out that decline in salmon stocks may result from problems other than habitat degradation. However, because wild salmon have such important cultural meaning within this region and because people have experienced seeing salmon returning from the sea to spawn in the streams of their origin, this indicator has both intellectual and emotional appeal, which translate into political support for measures to improve their return rates. Other examples of this effort to balance technical with popular concerns can be seen in Table 5.1, which presents the full list of the 40 topics addressed by this programme.

Planning theory assists both in suggesting the process for developing sustainability indicators, and their appropriate role in decision-making. Traditional definitions of planning hold that it is an exercise in purposeful, informed decision and action: an effort in functional, technical or applied rationality. In the context of planning for sustainable development however, we are often dealing with complex systems, in that our limited comprehension of how they work results in deep uncertainty, and in that there are many legitimate perspectives or views on any problem. This means that purely technically and scientifically grounded indicators, which can be specified by national governments and international organizations for application in most contexts, are less certain to be successful. That is, functional rationality as an approach has limited value for addressing issues of sustainability. Instead of taking an indicator as a one-to-one representation of reality or as a representation of an objective world, an indicator can be seen as a representation of what is felt important by the various actors, and as such is emotional representation made by the subject of the quality of the environment in which the subject is living.

Several other recent contributions to planning theory provide alternative models explaining planning practice and proposing alternative approaches. For example, Hiller (1995) argues for *phronesis* (or prudence) as an appropriate mode of thought in reaching complex decisions. As developed by Aristotle, phronesis requires candidness, reflection on values, prudent judgment, and persuasion to arrive at practical reason among parties contributing to the deliberation (Dryzek 1990). It does this in a manner sensitive to the particular context, and seeks to engage common sense, and experience (Hiller 1995). These are characteristics shared by substantive rationality, in contrast to the focus of functional rationality on the logic of scientific inquiry (Nozick 1993).

Two streams of current planning theory, *pragmatism* and *communicative action*, are informed by phronesis in that they embrace collective deliberation as the prime basis of decision-making. At the centre of pragmatism is the notion that the meaning of an idea or decision requires that we look at its consequences (Blanco 1994). While pragmatism holds that we can never have certain truth, we can use perception, reflection, and judgment employing social and intra-personal processes to gain understanding and to establish a common knowledge (Forester 1993). As such, pragmatism is a philosophy presenting tools to cope with what is referred to in Chapter 7 as *complicated* and *complex* issues, and actor consulting represents a pragmatic technique to resolve these planning problems.

Communicative action shares many ideas with pragmatism. Yet it is differentiated in its reduced emphasis on the ends-means concerns of pragmatism, and by its greater emphasis on interpersonal understanding, socialization of stakeholders, and co-ordination of actions through discussion (Innes 1995). Following Habermas (1984), communicative action also seeks to recognize power relations, and holds that the force of open and free argument should determine the decision reached by the group.

Consequently both the communicative action and pragmatism models for decision making offer procedural guidance in dealing effectively with sustainability and provide criteria for determining the nature of the knowledge which should inform these efforts. Thus, in addition to the desirability of developing and using indicators that the public understands and cares about, the very complexity of sustainability is the basis for a major argument supporting participation of actors in the process of formulating these measurements. As presented in this book, problems become increasingly complex as they involve relationships that are not understood, and goals on which there is no substantial agreement.

As Christensen (1985) notes, problems for which we have cause and effect understanding and which address widely held goals lend themselves to technical solution and routine decision making. However, if understanding of means for solving a problem is limited, or if there is disagreement concerning planning goals, then a simply technical approach to decision making is inappropriate. In these instances, a mixed strategy is required, where factual knowledge is important for informing decision making, but where reaching a decision is dependent on judgment as well. In a democracy, expert advice is normally limited to the factual elements of the decision, and affected parties are recognized as the valid sources of the judgment or value content of the decision (Sagoff 1988; Barnes 1999). In choosing and using sustainability indicators, the questions concerning what to measure, how to measure the social-economic-environmental effects of plans and programmes, and especially the task of balancing these dimensions of sustainability are all principally matters of judgment and thus call for broad participation by actors (Creighton 1993). This can be done by direct interaction (reflecting 'communicative action') as well as through actor-consulting (reflecting 'pragmatism').

Table 5.1   Indicators of Sustainable Community – Sustainable Seattle

---

Environment
- Wild salmon runs through local streams
- Biodiversity in the region
- Number of good air quality days per year, using Pollutant Standards Index
- Amount of topsoil lost in King County
- Acres of wetlands remaining in King County
- Percentage of Seattle streets meeting 'Pedestrian-Friendly' criteria

Population and Resources
- Total population of King County (with annual growth rate)
- Gallons of water consumed per capita
- Tons of solid waste generated and recycled per capita per year
- Vehicle miles travelled per capita and gasoline consumption per capita
- Renewable and non-renewable energy (in BTUs) consumed per capita
- Acres of land per capita for a range of land uses (residential, commercial, open, etc.)
- Amount of food grown in Washington, food exports, and food imports
- Emergency room use for non-emergency purposes

Economy
- Percentage of employment concentrated in the top ten employers
- Hours of paid employment at the average wage required to support basic needs
- Real unemployment, including discouraged workers, differentiated by ethnicity, gender
- Distribution of personal income, with differentiation by ethnicity and gender

- Average savings rate per household
- Reliance on renewable or local resources in the economy
- Percentage of children living in poverty
- Housing affordability gap
- Health care expenditures per capita

Culture and Society
- Percentage of infants born with low birth weight (including by ethnicity) ·
- Ethnic diversity of teaching staff in elementary and secondary schools
- Number of hours per week devoted to instruction in the arts for elementary, secondary schools
- Percent of parent / guardian population that is involved in school activities
- Juvenile crime rate
- Percent of youth participating in some form of community service
- Percent of enrolled 9th graders who graduate from high school (by ethnicity, income, gender)
- Percent of population voting in odd-year (local) primary elections
- Adult literacy rate
- Average number of neighbours the average citizen reports knowing by name
- Equitable treatment in the justice system
- Ratio of money spent on drug and alcohol prevention, treatment vs. incarceration for related crimes
- Percentage of population that gardens
- Usage rates for libraries and community centres
- Public participation in the arts
- Percent of adult population donating time to community service
- Individual sense of well being

5.5 Criteria for use in designing sustainability indicators

So far, we have discussed goals of sustainability, the purposes and uses of sustainability indicators, and the outline of alternative processes for developing these indicators. This leads to our considering criteria to be used in designing a set of sustainability indicators, and in critically assessing the results. This design process starts with identifying or enumerating the desirable features of sustainable development, and then devising measurements for each of these features (Rees 1998). While an exhaustive list of criteria for use in this design process could be very long (Maclaren 1996), the crucial elements are summarized in the following list of guiding principles for preparing a set of sustainability indicators. In parallel to this approach, Chapter 8 explores the use of an actor-consulting model as a communicative means to identify realistic and meaningful objectives and indicators, which include those related to sustainability, at the various stages of the planning cycle. This focuses on the perceptions, motives, interests and concerns of parties involved and to aim for *the exploitation of the knowledge* and for *an understanding of the value judgements* of the actors.

Valid   It is especially important that indicators accurately represent the sustainability features that the planning seeks to address; that they relate to desired ends (Constanza 1993; Cobb et al. 1995). In this connection, both positive and negative impacts of human activity need to be included. From both a communicative and actor-consulting perspective validity of indicators has to be found not only on technical grounds (since these are easily disputable) but also on the basis of arguments contributing to consensus or supporting mutual understanding.

Measurable   Indicators need to be specified in operational terms. In a traditional and functional mode of reasoning, this means showing not only the direction but also the rate of change affecting social, economic and environmental well-being. Consequently, it is most useful if the measurements employ interval level data, and if they reflect cause-effect relationships (Turner and Tschirhart 1999). In a communicative mode of reasoning, criteria should address conditions under which processes of planning are or can be considered. Criteria should also focus on (changes of) the various actors their interpretations, motives, actions and roles.

Understandable   Indicators need to be expressed in terms that are widely intelligible and useful to decision making (Milbrath 1989). They should be designed with their audiences in mind, and presented in simple and clear language. Graphics are useful to assist communication. Technical terms are likely to be confusing unless defined and illustrated.

Focus on principal concerns   Characteristics of the urban (or rural) system for which sustainability indicators are selected need to be closely related to issues about which the public cares or can come to care about. Linking these items to broad public experience can contribute to understanding and political support (Jacobs 1997). Employing a large number of indicators can be confusing; grouping and summarizing them under a limited set of categories will assist the intended audience to make sense of them. The extent to which the selected set of indicators adequately addresses the complexity of relevant sustainability issues should be explicit.

Use carefully selected sources   Since primary-source or new evidence concerning measured performance is expensive to collect, it is attractive to use secondary-source data where possible. The reliability of data needs to be assessed, assumptions and uncertainties made explicit, and the sources of the data should be made clear (Pearch 1993).

Use standardized measures   In a technical-rational approach to planning, a principal value of indicators is in comparing performance with reference points such as meeting specified targets. Employing measures that are widely used by others allows comparisons with other communities and with historical data (Mol 1999). Standardization of several indicators permits comparison of rates of change over these variables. Basically the same argument counts for indicators used within a communicative mode of planning, addressing conditions of planning processes and addressing the actors involved, however there should be agreement among these actors about the indicators used.

Seek output or outcome measures Since the principal use of sustainability indicators is to measure changes in a system, measures of effects are preferred over measures of input (Jackson and Marks 1994). Similarly, using measures of conditions sought helps in avoiding prematurely specifying which course of action is thought to be the most effective.

Determine spatial and time dimensions Since many of the social, economic and environmental features of interest in planning for sustainable development are characteristics of regions larger than the neighbourhood or municipal context, the spatial base for this information should recognise the regional context (Sawicki and Flynn 1996). Similarly, indicators should be designed and used for repeated measurements over time, in order to identify trends and to support review and revision of plans and of the goals on which they are based. An iterative process of this sort takes advantage of new insights and priorities, thus accommodating uncertainty, and also supports learning by doing (Webler et al. 1995).

Facilitate public interaction As noted earlier, useful indicators normally balance fact and judgment. They may rely on technical sources for factual content, with the affected parties having the principal role in providing the value content. This consultation and participation should ensure that a range of views is accommodated, and that an informed constituency for the set of measures is developed (Hajer 1995). Public familiarity with these measures commonly results in their being used in political dialogue and for ensuring public agencies and other parties are accountable (Webler 1999).

Clear assignment of responsibility Success in designing and using a set of sustainability indicators requires agreements to be reached concerning the roles of the various participants. Agreement is needed for example on timescales for the completion of data collection and reporting (El Serafy 1997). Realistic estimates of the resources required to complete these tasks, and establishing commitments for support are important features of a management programme.

These ten criteria for designing an effective set of sustainability indicators illustrate the complexity of putting together a programme to accomplish this task. Specific criteria may vary in importance depending on the institutional and physical context and the complexity of the issue at hand, and on how the particular set of measures will be used (Healey and Williams 1993). As features of a successful exercise to specify and employ sustainability measures, these criteria are both useful in guiding how to go about doing this, and in assessing in retrospect the effectiveness – the likely sources of success and shortcomings – of past efforts.

5.6   Conclusions

While sustainability is widely embraced as a new paradigm of urban planning, it is also increasingly accepted as a fuzzy notion. There is growing agreement concerning the abstract outlines of what constitutes sustainability – the balancing of economic, environmental and social objectives in a manner that addresses a long-term time horizon. Rather confusing however are the many views concerning how these dimensions should be defined and balanced (Ravetz 2000). This results in individual perceptions of how to behave and act in a 'sustainable' way, which can have disastrous effects when sustainability has to be made operational within a planning process. While there may be agreement on the desirability of 'sustainable planning' in general terms, the implementation of these ideals might fail to produce a positive outcome, due to counterproductive interpretations. Design and use of a set of sustainability indicators requires somehow reconciling these views and, importantly, translating them into operational terms, which may be applied in planning, decision-making, and management. As described in this chapter, planning, monitoring, stock-taking, and management all depend for evidence and direction on metrics which facilitate comparison of options and tracking of progress. The functional criteria illustrated in this chapter can be used to design a set of sustainability indicators that meets both substantive and procedural demands, balancing validity with political purposes.

As sustainability is a fuzzy notion, sustainability-related indicators should be the result of argumentation, the balancing of thoughts and views and the outcome of consensus. In other words indicators should be identified through exploitation of the knowledge of crucial parties. In this chapter, two theoretical approaches have been presented as a means of identifying these indicators. One is an approach inspired by *communicative action*, usually leading to direct interaction between the parties involved. The other is inspired by *pragmatism*, a philosophy dealing with the interface between complicated and complex issues (see Chapter 7), with actor-consulting as a possible tool to guide the process. In the latter case parties need to be consulted about their perception of sustainability, and how they consider results and progress might be measured. In that sense indicators can be seen as a means to stimulate the actors' ideas about how planning should develop at the various stages of the planning process.

When reflecting on the argumentation brought forward in this chapter about indicators, we are now able to identify two basic functions, which relate closely to the process of actor-consulting. One is that indicators can only be effective if they relate to people's *perception* of how the indicators reflect the issue at hand – here sustainability. In this sense, indicators represent an *output* from the actor-consulting model. The other is that indicators can also represent information extracted from the planning process, and which can then be used as a means to examine the reflections and argumentations of a variety of stakeholders (*input* to the actor-consulting model). As shown in figure 8.6 (Chapter 8), indicators can thus be utilized as both an *input* to, and an *output* from, the actor-consulting model.

References

Abramovitz, J. (1997) 'Valuing Nature's Services', in: L. Brown, C. Flavin and H. French (eds), *State of the World*, Worldwatch Institute, Washington, D.C., pp. 95–114.

Barnes, M. (1999) 'Researching Public Participation', *Local Government Studies*, Vol. 25, pp. 60–75.

Bartelmus, P. (1999) 'Green Accounting for a Sustainable Economy: Policy Use and Analysis of Environmental Accounts in the Philippines', *Ecological Economics*, Vol. 29, pp. 155–170.

Beatley, T. and K. Manning (1997) *The Ecology of Place: Planning for Environment, Economy, and Community*, Island Press, Washington, D.C.

Berke, P. and M.M. Conroy (2000) 'Are We Planning for Sustainable Development? An Evaluation of 30 Comprehensive Plans', *Journal of the American Planning Association*, Vol. 66(1), pp. 21–33.

Blanco, H. (1994) *How to Think About Social Problems: American Pragmatism and the Idea of Planning*, Greenwood Press, Westport (US).

Burby, R.J. and P.J. May (1998) 'Intergovernmental Environmental Planning: Addressing the Commitment Conundrum', *Journal of Environmental Planning and Management*, Vol. 41(1), pp. 95–110.

Campbell, S. (1996) 'Green Cities, Growing Cities, Just Cities? Urban Planning and the Contradictions of Sustainable Development', *Journal of the American Planning Association*, Vol. 62(2), pp. 296–312.

Christensen, K.S. (1985) 'Coping With Uncertainty in Planning', *Journal of the American Planning Association*, Vol. 51(1), pp. 63–73.

Cobb, C., T. Halstead and J. Rowe (1995) *The Genuine Progress Indicator – Summary of Data and Methodology*, Redefining Progress, San Francisco.

Common, M. (1995) *Sustainability and Policy*, Cambridge University Press, Cambridge (UK).

Corson, W. (1995) *An Inventory of Local and Regional Programs Using Goals and Indicators to Define and Measure Quality of Life, Progress, and Sustainability at the City, County, and State Level*, Global Tomorrow Coalition, Washington, D.C.

Costanza, R. (1993) 'Developing Ecological Research that is Relevant for Achieving Susstainability', *Ecological Applications*, Vol. 3, pp. 579–581.

Creighton, J. (1993) *Guidelines for Establishing Citizens' Advisory Groups*, U.S, Department of Energy, Washington, D.C.

Daniels, S.E. and G.B. Walker (1996) 'Collaborative Learning: Improving Public Deliberation in Ecosystem-Based Management', *Environmental Impact Assessment Review*, Vol. 16, pp. 71–102.

Davies, H.W.E. (1989) *Planning Control in Western Europe*, HMSO, London.

Deyle, R.E. and R.A. Smith (1998) 'Local Government Compliance with State Planning Mandates: The Effects of State Implementation in Florida', *Journal of the American Planning Association*, Vol. 64(4), pp. 457–469.

Dryzek, J. (1990) *Discursive Democracy: Politics, Policy and Political Science*, Cambridge University Press, Cambridge.

El Serafy, S. (1997) 'Green Accounting and Economic Policy', *Ecological Economics*, Vol. 21, pp. 217–229.

Environment Agency (1998) *Consensus – Building for Sustainable Development*, Environment Agency, London.

Forester, J. (1993) *Critical Theory, Public Policy and Planning Practice: Towards a Critical Pragmatism*, State University of New York Press, Albany (US).

Freeman, A.M. (1993) *Measurement of Environmental and Resource Values: Theory and Methods*, Resources for the Future, Washington, D.C.

Freilich, R.N. (1999) *From Sprawl to Smart Growth*, American Bar Association, Chicago.

Gouldson, A. and J. Murphy (1997) 'Ecological Modernization: Restructuring Industrial Economics', in: M. Jacobs (ed.) *Greening the Millennium? The New Politics of the Environment*, Blackwell, Oxford, pp. 74–85.

Habermas, J. (1984) *The Theory of Communicative Action*, Beacon Press, Boston.

Hajer, M. (1995) *The Politics of Environmental Discourse: Ecological Modernization and the Policy Process*, Clarendon Press, Oxford.

Hart, M. (1995) *Guide to Sustainable Community Indicators, QLF / Atlantic Center for the Environment*, Ipswich (US).

Haughton, G. and C. Hunter (1994) *Sustainable Cities*, Jessica Kingsley, London.

Healey, P. and R. Williams (1993) 'European Urban Planning Systems: Diversity and Convergence', *Urban Studies*, Vol. 30, pp. 699–718.

Hiller, J. (1995) The Unwritten Law of Planning Theory: Common Sense', *Journal of Planning Education and Research*, Vol. 14(4), pp. 292–296.

Howe, D.A. (1993) 'Growth Management in Oregon', in: J. Stein (ed.), *Growth Management: The Planning Challenge of the 1990s*, Sage, Newbury Park (US), pp. 61–75.

Innes, J.E. (1995) 'Planning Theory's Emerging Paradigm: Communicative Action and Interactive Practice', *Journal of Planning Education and Research*, Vol. 14(3), pp. 183–189.

Jackson, T. and N. Marks (1994) *Measuring Sustainable Economic Welfare*, Environment Institute and New Economics Foundation, Stockholm.

Jacobs, M. (1997) 'Introduction: The New Politics of the Environment', in: M. Jacobs (ed.), *Greening the Millennium? The New Politics of the Environment*, Blackwell, Oxford, pp. 1–17.

Kaiser, E., D. Godschalk and F. Chapin (1995) *Urban Land Use Planning, Fourth Edition*, University of Illinois Press, Urbana and Chicago.

Lafferty, W.M. and J. Meadowcroft (eds.) (2000) *Implementing Sustainable Development: Strategies and Initiatives in High Consumption Societies*, Oxford University Press, Oxford.

Limoges, C. (1993) 'Expert Knowledge and Decision-Making in Controversy Contexts', *Public Understanding of Science*, Vol. 2, pp. 417–426.

Logan, J.R. and H.L. Molotch (1987) *Urban Fortunes: The Political Economy of Place*, University of California Press, Berkeley (US).

Lynn, F.M. and J. Kartez (1995) 'The Redemption of Citizen Advisory Committees: A Perspective From Critical Theory', in: O. Renn, T. Webler and P. Wiedemann (eds), *Fairness and Competence in Citizen Participation: Evaluating Models for Environmental Discourse*, Kluwer Academic, Boston, pp. 87–102.

Lynn, F.M. and G.J. Busenberg (1995) 'Citizen Advisory Committees and Environmental Policy: What We Know, What's Left to Discover', *Risk Analysis*, Vol. 15(2), pp. 147–162.

McDonald, G.T. (1996) 'Planning as Sustainable Development', *Journal of Planning Education and Research*, Vol. 15, pp. 225–236.

Margerum, R.D. and S.M. Born (1995) 'Integrated Environmental Management: Moving from Theory to Practice', *Journal of Environmental Planning and Management*, Vol. 38(3), pp. 371–391.

Milbrath, L. (1989) *Envisioning a Sustainable Society: Learning Our Way Out*, State University of New York Press, Albany (US).

Miller, D. (1997) 'Dutch Integrated Environmental Zoning: A Comprehensive Program for Dealing With Environmental Spillovers', in: D. Miller and G. de Roo (eds), *Urban Environmental Planning*, Avebury, Aldershot (UK), pp. 165–178.

Mol, A.P. (1999) 'Ecological Modernization and the Environmental Transition of Europe: Between National Variations and Common Denominators', *Journal of Environmental Policy and Planning*, Vol. 1(2), pp. 167–181.

National Research Council, Board on Sustainable Development (1999) *Our Common Journey: A Transition Toward Sustainability*, National Academy Press, Washington, D.C.

Nelson, A.C. (1986) 'Using Land Markets to Evaluate Urban Containment Programs', *Journal of the American Planning Association*, Vol. 52(2), pp. 156–171.

Nelson, A.C., J.B. Duncan (1995) *Growth Management Principles and Practices*, Planners Press, Chicago.

Nozick, R. (1993) *The Nature of Rationality*, Princeton University Press, Princeton (US).

Pearce, D.W. (1993) *Blueprint 3: Measuring Sustainable Development*, Earthscan, London.

Petts, J. (1992) 'The Public – Expert Interface in Local Waste Management Decisions: Expertise, Credibility and Process', *Public Understanding of Science*, Vol. 6(4), pp. 359–382.

Ravetz, J. (2000) *City – Region 2020 Integrated Planning for a Sustainable Environment*, Earthscan, London.

Reardon, K.M. and T.P. Shields (1997) 'Promoting Sustainable Community/ University Partnerships Through Participatory Action Research', *National Society for Experimental Education Quarterly*, Vol. 23(1), pp. 22–25.

Rees, W.E. (1998) 'Understanding Sustainable Development', in: B. Hamm and P.K. Muttagi (eds) *Sustainable Development and the Future of Cities*, Intermediate Technology, London.

Sagoff, M. (1988) *The Economy of the Earth: Philosophy, Law and the Environment*, Cambridge University Press, Cambridge (UK).

Sawicki, D.S. and P. Flynn (1996) 'Neighborhood Indicators', *Journal of the American Planning Association*, Vol. 62(2), pp. 165–183.

Selman, P. (1998) 'Local Agenda 21 – Substance or Spin?', *Journal of Environmental Planning and Management*, Vol. 41(5), pp. 533–553.

Smith, J.F. (1998) 'Does Decentralization Matter in Environmental Management?', *Environmental Management*, Vol. 22, pp. 263–276.

Sustainable Seattle (1993) *Indicators of Sustainable Community – A Report to Citizens on Long-Term Trends in Our Community*, Sustainable Seattle, Seattle (US).

Taylor, D. (1991) *Sustaining Development or Developing Sustainability? Two Competing World Views*, Alternatives, New York.

Troy, P.N. (1996) 'Environmental Stress and Urban Policy', in: M. Jenks, E. Burton and K. Williams (eds) *The Compact City: A Sustainable Urban Form*, Spon, London, pp. 200–211.

Turner, P. and J. Tschirhart (1999) 'Green Accounting and the Welfare Gap', *Ecological Economics*, Vol. 30, pp. 161–175.

Webler, T., H. Kastenholz and O. Renn (1995) 'Public Participation in Impact Assessment: A Social Learning Perspective', *Environmental Impact Assessment Review*, Vol.15, pp. 443–464.

Webler, T. (1999) 'The Craft and Theory of Public Participation: A Dialectical Process', *Journal of Risk Research*, Vol. 2(1), pp. 55–72.

World Commission on Environment and Development (1987) *Our Common Future*, Oxford University Press, London.

Yap, N. (1989–1990) 'NGOs and Sustainable Development', *International Journal* XLV, pp. 75–105.

Part B
The Actor-Consulting Model

Chapter 6

Shifts in Planning Practice and Theory: From a Functional Towards a Communicative Rationale

Gert de Roo[1]

6.1 Coping with fuzziness

> Beware of beautiful days. Bad things happen on beautiful days. It may be that when you get happy, you get careless. Beware of having a plan. Your gaze is focused on the plan and that's the moment when things start happening just outside your range of vision (French 2002: 1).

Planners and decision-makers managing our physical environment in line with societal needs all know that the adoption of a plan or the ratification of a decision is just the beginning, even if there is a strong consensus among all the participants involved. Even when there is consensus, plans and decisions still end quite regularly in either a lack of action, or – when situations get totally out of hand – in unwanted, negative and undesired outcomes.

For a long time planners assumed that controlling the physical environment on the basis of technical, instrumental and procedural expertise was the way to go (see Friedmann 1987, Meyerson and Banfield 1955). Absolute control, as founded on theoretical grounds, has however led to outcomes that are infeasible and impractical, particularly in an increasingly democratic and equitable society. Planning systems that feature a high degree of top-down control in many ways represent the remains of a system based on nineteenth century political ideals, and on post-war functionalism. Planners have nevertheless continued to seek certainty, and in doing that they have adjusted their approaches in various pragmatic ways (see among others Etzioni 1968, Lindblom 1959, March and Simon 1958, Rittel and Webber 1973, Simon 1960). The option to leave control – or the responsibility for it – to others and to accept uncertainty as an inseparable part of the real world has hardly been a consideration.

In the previous chapters it has been made very clear that certainty and control are perhaps not the only way forward. This challenges traditional thinking in planning, where fuzziness or fluidity of notions, encompassing a wide range of perceptions

---

1   Gert de Roo is Professor in Planning at the Department of Planning and Environment, Faculty of Spatial Planning at the University of Groningen, Groningen, The Netherlands.

and motivations, as well as unexpected behaviour and interpretations, have been considered as barriers to the performance of planning processes. Even so it is often unclear on what level the decision-making responsibility should to be taken and who has to take the lead in solving the issue. And after that has been settled it remains to be seen what approach has to be selected to solve the matter. This contributes to uncertainty, and often results in planning processes that are faced with a high degree of trial-and-error.

In this book we question whether we should be bound by 'trial-and-error'. Is there an escape route that brings us to safer grounds? In this chapter (§ 6.2) we will first of all review the arguments made in the previous chapters. This should allow us a breathing space to reflect on the key issues. Paramount among these is our growing realisation that within the planning domain there are aspects that can be seen as fuzzy, fluid, illusive or unclear, despite our first impressions and in contrast with our predispositions. If we are willing to recognise these fuzzy elements, we have made an important step forward in coping with fuzziness in planning.

In order to develop our sense of perspective, section 6.3 will carry out a review of post-war European developments in planning practice. This is complemented in Section 6.4 with a review of the discussions and developments that have taken place in recent years in the field of planning theory. Both sections will show how thinking has been led – in particular at the beginning – by a search for control and – later on – by a search for strategies for coping with uncertainty. The arguments made in this chapter will be the input to Chapter 7, which tries to understand and to cope with the fuzzy nature of planning.

6.2   Reflection on Part A

Let us firstly review our progress so far. In Chapter 2 Healey has taken the lead in the discussion about fluidity and fuzziness in planning, by examining the difficulties inherent in grasping the divergent understandings and meanings of some important concepts. The key dimensions of 'sustainability' and 'sustainable development' are used as examples. Many, she points out, have questioned the extent to which the new sustainability rhetoric has become 'reality'. The rhetoric and the ideas behind it should naturally lead to a transformation of government institutions and policy processes, but instead the sustainability discourse tends to be channelled into established policy frameworks, and the outcome is to dilute the implementation of policy intentions. Struggles about frames of reference (how to think about an issue), and struggles about who should be involved in policy design and implementation form a barrier to attempts to give sustainability operational meaning. This has led in many cases to unrealistic expectations for the outcome of 'sustainable development' policies. Accordingly, we find increasing pressure for the transformation of the institutions that implement contemporary notions such as 'sustainability'. Techniques are needed to identify the role of actors in contributing to this transformation.

Martens pursues this discourse in Chapter 3, beginning with the observation that actors often base their expectations and actions on implicit frames of reference, without questioning them. Martens argues that this situation is largely the outcome of the adoption in many countries of formal planning and policy systems that reflect the coordinative model of governance, having a single authority, which collects information, sets goals and priorities, and selects and implements policies. These formal planning systems have enabled planning authorities to 'force' the various actors into a role that is convenient for the prevailing institutional system. This situation has given rise to steady pressure for institutional change, resulting in hybrid governance systems that include different combinations of the various features of the coordinative, competitive and communicative models of governance. These integrated models are dubbed 'fuzzy' models of governance, because they are not based on any clear-cut model. This has already led to an increasing variation in the roles and responsibilities of actors. Instead of a clear, rigid and top down responsibility for each area of policy, an approach based on shared responsibilities by the various actors involved, is becoming common within the planning arena. These transformations mean a shift from an object-oriented form of planning towards an inter-subjective approach, where the roles, perceptions, behaviour and motivations of the actors involved are becoming increasingly important.

This transformation in governance does not only affect all actors involved in the planning process. It also impacts upon the formal planning and policy systems. In Chapter 4, Voogd and Woltjer illustrate the complex structures of planning activities taking place at the regional level of government, defined here as the level between local and national government. The transitions currently taking place in the Netherlands' provincial level are used as an example. Here, the responsibility to involve actors in the planning process is still seen as a burden rather then a challenge, although the authorities face a growing call for shared responsibility. This adds to the fuzziness of regional governance and planning. Voogt and Woltjer argue that regional planning should be supportive of a consensus planning process involving a wide range of actors, as a means to deliver for example quality in development planning at the local level, whilst retaining an element of top-down government. In this environment, coordinative techniques that help to develop an understanding of the various frames of reference of actors are likely to be of significant value.

Miller explores a different angle. While the authors so far have focussed on policy, action and interaction, Miller looks at *information* that should be made available for decision-making purposes. Clearly, as our institutional systems move away from the coordinative model towards more fuzzy models of governance, the processes by which information is gathered, presented and used must also change. Miller uses 'sustainability' as an example. Criteria are suggested for designing indicators, as a means to guide and inform policy processes that are meaningful to the actors involved. A major use of sustainability indicators is to translate abstract statements of goals or visions into concrete, operational terms. Miller points out that indicators should be seen as a means of communicating information to stakeholders concerning changes that are taking place. Indicators can no longer be seen solely from a traditional perspective, as they represent – effectively – the information that feeds arguments into a discussion or 'learning process'. Quantitative and object oriented indicators will act alongside, will be complemented by or will be replaced by qualitative and communicative indicators to represent both what is going on and what is felt to be important. Development and use of measures of this nature can become a focus for action *and* interaction. In that sense indicators become a communicative tool, rather than – traditionally – a tool to measure whether targets are met.

Let us now attempt to synthesise some of the arguments put forward in these opening chapters. Healey sees that the divergent understandings and intentions of the various actors have the tendency to form a new and fuzzy reality represented by a multiplicity of frames of reference. This suggests a need for a better understanding of notions, concepts, doctrines, plans, decisions, and the way these relate to and interact with the institutional setting. Martens foresees a shift in attention from object oriented planning towards a form of planning based on the role of actors. Voogd and Woltjer support the need for an effective intermediate level of governance, which must become more open, active and interactive.

So in summary, we see that from different perspectives, Healey, Martens and Voogd & Woltjer foresee a transformation of models and systems of governance, moving from a traditional top-down form towards a pluralistic governance system, which adapts in accordance with the balance of the various interests, and the relations between stakeholders. Policy control varies, depending on the specifics of the situation at hand. Miller adds to the discussion the role of information in general and of indicators in particular. These can be seen as a function for communication in a modern society in which no longer one single body is responsible for making decisions, but numerous actors, parties and stakeholders are involved. From a theoretical perspective – as we shall shortly see – this represents a shift from a technical rational towards a communicative rational approach. It means that the 'how' question, which was traditionally the key question in planning, is being superseded by the 'why and wherefore' questions and that the latter are moving to the centre stage of the planning arena.

6.3   Historical reflection: The end of planning control?

Transitions in our society, and in particular in our systems of governance continue to capture our interest. However, it is perhaps more important to pinpoint the motives that have led to these transitions. In Europe the predominant governance system of the last forty to fifty years has been the so-called coordinative model of governance. The main feature of this model has been a formal planning and policy system with a top-down structure, which expects actors to express solidarity with the system's preferences. In that sense the government is in 'full control', based on the assumption that it knows what is good for us all. It expects lower authorities to *perform* according to the decisions that it makes and expects citizens to *conform* to these decisions.

In most countries of north-west Europe, despite an impressive history of planning practice, physical planning can still be seen as a struggle, where trial and error predominates. Many planning authorities are experiencing difficulties in shaping their environment according to their wishes. Instead of motivating and guiding developments, too often it seems that developments are taking their own course, with planning and planners – again – one step behind in adapting to reality. Although the coordinative model of governance has always displayed these imperfections, the evidence suggests that our inclination is to try to adjust to the prevailing system, rather than to challenge it with a more radical approach to governance. Nevertheless, this prevailing discourse in planning and policy-making has been challenged over the years, and lately with some success.

Let us examine the motives that have driven the planning systems of Western Europe during the second half of the twentieth century. This will lead us to a series of planning models that we can then go on to explore in greater detail. The end of World War II showed a continental Europe in ruins. German occupation had left a rather rigid but well equipped planning system in most occupied countries, with three interacting layers of policy-making. The UK initially pursued a similar three-tier system. Germany itself was forced to adopt a 'Lander' system, incorporating a fourth layer, subdividing the nation into more-or-less autonomous regions. In most countries the policy system was based on what is termed delegated responsibility. This was intended to be a decentralised system of executing national guidelines within a centrally controlled policy framework. National governments were to design a strategy and the regional and local authorities were to implement the strategy through their plans. Due to the extenuating circumstances, the beginning of this post-war planning era was inevitably functional in its nature.

'Rational' planning

The start of the post-war planning era in north-west Europe is seen by many as 'rational'. This type of planning 'demands the systematic considerations and evaluation of alternative means in the light of the preferred ends they are to achieve" (Alexander 1984; 63). In this instance, rationality can be defined as a method 'to have criteria of success laid down in advance" (Sagoff 1988). Nowadays this type of rationality is referred to as 'instrumental' (Dryzek 1990), 'technical' (Healey 1983) or 'procedural' (Faludi 1987). In this sense the north-west European planning rationale was indeed technical, instrumental, procedural and – in other words – functional. The cooperative model of governance was well established, including the assumption of government control. In fact this approach matched very well the expectations of society with its military sense of obedience following the years of the War.

In the 1960s, quality, and that of the local environment in particular, became important in all western European countries. Although Germany and the Netherlands had not yet solved their housing shortages, the functional approach was no longer perceived as the only option. By embracing *scenario* approaches, another step in planning was made by composing spatial visions, which gave the process of planning a focus towards the future. Although spatial policy remained rather functional, it also differentiated between each of the country's distinct regions. This went hand in hand with a growing confidence that physical developments were controllable (Davies 1989).

Sectoral specialisation of planning

In the 1970s, several incidents served to put the trust in a controllable society under pressure. Throughout Europe, the economic situation experienced a dramatic decline as a direct result of the Middle East Oil Crisis, and the onset of the collapse of the Fordist production model. Population forecasts were put to the test. Authoritarian decisions were no longer taken for granted. Citizens became increasingly critical, and organised themselves into interest groups. At the same time government policy in countries such as France, Germany, Denmark, Sweden and the Netherlands rapidly expanded into several sectors, in an effort to keep in control of the many internal and external developments confronting the physical environment. In addition to spatial planning, water management, traffic and transport policy, and planning of the grey and green environment emerged as distinct fields in planning.

The 1980s witnessed a substantial elaboration of these policy sectors as each developed its own legal system, planning system, specialised instruments, financial structure, and professional organisation, including formal and informal networks. These became highly specialised, including the development of sector-specific languages. The outcome was a sharply divided planning system, based on several strong sectors, each claiming authority over their peers. The result of this specialisation of policy and physical planning was that policy-making came to have little to do with 'controlling' the outside world through planning, as each government department struggled to further the extent of its influence and control.

Losing control?

The result of specialisation became noticeable in the 1990s as dilemmas between the policy sectors emerged. One of the more prominent dilemmas was the so-called 'paradox of the compact city' (Bartelds and De Roo 1995, Breheny 1996, De Roo 2000, Jenks, Burton and Williams 1996). As discussed in Chapter 1, the compact city was originally a spatial concept that was adopted throughout Europe as a panacea for promoting social, economic and spatial advantages within urban settings, and for relieving the countryside from urban pressure. Local authorities in Europe initially adopted this policy with enthusiasm. The Netherlands was among the first countries to identify the contradictions of the 'compact city' in its quest to transform the policy into practice (Bartelds and De Roo 1995). Simultaneously with the compact city policy, environmental policy in the Netherlands matured quickly on the basis of a rapidly expanding environmental zoning programme (De Roo 1993). This programme was meant to cope with environmental externalities emanating from industry and traffic, including noise, odour, dust, emissions of toxic substances, industrial disasters and so on. The main principle was to translate environmental standards into spatial zones, expressing the distance between environmentally intrusive activities, such as industry and traffic, and environmentally sensitive functions, in particular housing and residential areas. Maintaining distance between functions was however generally incompatible with the compact city concept. The outcome was a clash between spatial planning and environmental policy in the Netherlands (De Roo and Miller 1997). This is one example of many policy conflicts that arise out of an expanding and highly specialised policy system that is unable to foster cohesion within the system itself, and thus remains inconsistent in approaching the physical environment.

The post war decades have therefore followed a continuing series of responses and adaptations to the rational planning approach where above all, we have attempted to stay in control as much as possible. This behaviour has resulted in planning professionals trying to cope with highly specialised planning systems, unable to work with the 'compact city' concept and to plan in a 'sustainable' way. Most countries are continuing in their quest to control their physical environment by modifying and expanding their planning systems, still hoping to stay in control, somehow...

Responding to growing complexity

However, the 1990s did begin to show some interesting changes and shifts in planning and policy-making, reflecting the move towards 'fuzzy' models of governance. Slowly but surely, we have witnessed the integration of elements of the competitive and the communicative models into the coordinative model of governance. For example, the influence of market processes and public-private partnerships have risen substantially during the 1990s. Even more important has been the rise of communicative approaches in planning (see among others Healey 1996, Innes 1995, Sager 1994 and Woltjer 2000). Also the call for subsidiarity – the need for flexibility in respect of local or regional interpretation of rules and regulations, and to handle issues on the level at which they occur – has been embraced in various countries in Europe, resulting in a variety of decentralisation processes. In France in particular decentralisation has become an important issue. In the UK regional planning is being reinvented. In Belgium and the Netherlands area-specific policies are being introduced (De Roo and Schwartz 2001, Van den Broeck 2001). In Germany integration processes are becoming important (Greiving 2001).

All these initiatives are somehow interconnected, as they are a response to the implicit conclusion that one single entity – the national government – does not have the resources to control the physical environment in such a way that it satisfies all parties. More and more, issues need to be solved in accordance with the local or regional context. Hence, the desire for decentralised, issue related, area-specific policy, which is conducive to the participation of local actors and to integration of the relevant policy sectors. It also means at last a change away from traditional coordinative and functional approaches in planning and policy-making. This transformation should be seen as a response to the growing dynamics, complexity and uncertainties of our societies.

6.4   Theoretical reflections

The responses of planning systems to the stimuli of change are in themselves the subject of theoretical debate. We therefore turn to planning theory, in an attempt to gain a better understanding of how to deal with the growing complexity of our society and the consequent uncertainties that accompany this complexity.

In post-war Europe both the theory and practice of planning have developed continuously and simultaneously. Seen from a theoretical perspective, there was a strong demand for a technical-rationality approach, just after the Second World War. Europe had to be rebuild, and quickly. Obviously there was a need for certainty and control in planning, and technical rationality was the most desirable option (see Figure 6.1). This approach was based on the modernistic assumption that given all the information at hand at the beginning of a planning process, a clear outcome could be defined, and the final results would be fully predictable. Unfortunately, this only proved to be so in rare cases. These were the simplest ones with reliable cause-and-effect relationships. Nowadays, this period is regarded a time of primitive optimism (Voogd 1995). It is no surprise that this type of planning – blueprint planning – was susceptible to criticism

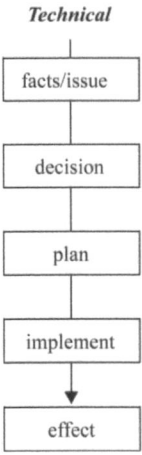

Figure 6.1   Technical planning processes

During the 1970s and 1980s various new constructions and explanations in planning theory became common, such as 'bounded rationality' by Simon (1960), 'mixed-scan' by Etziony (1967), the 'strategic choice approach' by Friend and Jessop (1969), and Lindblom's 'science of muddling through' (1959). In a way these constructions can be seen as responses to the shortcomings of the technical-rational approach. This period also brought us the scenario planning approach (see Figure 6.2), where certain information is taken as given at the beginning of the planning process, then various routes are constructed to tackle a spatial problem. Out of these possible routes, the most favoured is chosen, and maintained as long as realistically possible. Feedback loops can be introduced to maintain a realistic course, resulting in a model showing planning as a cyclical process. This proved to be a useful strategy for a variety of planning issues.

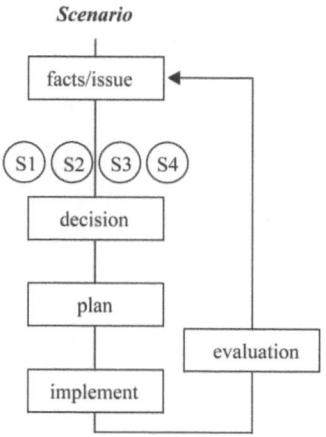

Figure 6.2   Scenario planning processes

However, there was no denying that for a substantial proportion of planning issues, in particular those involving many different actors, the scenario approach appeared to fall short as well. As a reaction to this situation, planners looked for other ways to solve their most difficult planning issues. The 'communicative rationale' approach was embraced as an alternative (see Figure 6.3).

Contrary to the theory supporting the technical-rational approach, communicative theory does not see planning issues as 'realities' in the outside world (see Chapter 2). They are seen as the abstract constructions of the various people involved. Through interaction among crucial actors, each actor's difficulties are evaluated, which results in a common understanding towards a commonly 'constructed' issue, a collective strategy on how to deal with the issue, and a plan that turns this strategy into action. This also holds for land use planning decisions. In contemporary planning practice many groups are involved and they influence the policy agenda and the outcomes of planning processes by means of their competence, status, legitimacy, knowledge, information, and money. Since actors tend to behave strategically, their goals can rapidly change over time and depend heavily on the planning context (Teisman 1995).

One might conclude that governments can not solely determine the public interest, let alone those of the individual parties involved (see Chapter 4). Consequently, the importance of communicative interaction has been acknowledged in contemporary planning practice, and communicative planning theories have become established in planning theory (see e.g. Innes 1995). However, even this approach has its limitations. A communicative strategy is only beneficial for those issues, where numerous actors have relatively equally-valued and mutually dependent, however opposing interests.

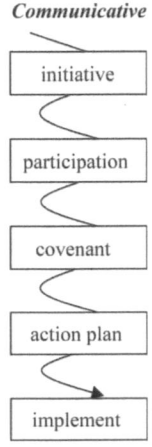

Figure 6.3   The communicative/participatory planning process

6.5   Concluding remarks

Accepting uncertainty as a reality requires fundamental changes in our belief system or frame of reference. Accepting uncertainty gives us an interesting way out of the planner's dilemma of how to cope with complexity and dynamics. Increasing uncertainty in planning necessitates a shift from a functional rational approach towards a communicative rational approach. To put it in a different context: as the complexity of a given issue increases, the approach in planning changes from a focus on the planning's object towards an inter-subjective or institutional perspective. Instead of maximising the planning result in line with the predefined goals, the focus will shift towards optimising the planning process. In the latter case it means that planners have become managers of change, in a continuously transforming environment.

If we can come to terms with a frame of reference that accepts uncertainty, we will recognise the restrictions that we have placed upon ourselves by adhering to the tradition of rational planning. This represents a significant step forward. It allows us room for new lines of reasoning that can enhance our planning concepts, can provide an insight into improved planning techniques, and can help us to confront the reality of the twenty-first century.

References

Alexander, E.R. (1984) 'After Rationalism, What?; A review of responses to paradigm breakdown', *Journal of the American Planning Association*, Winter, pp. 62–69.

Bartelds, H.J. and G. de Roo (1995) *Dilemma's van de compacte stad: uitdagingen voor het beleid* [Dilemmas of the compact city: challenges for policy-making], VUGA Uitgeverij, The Haque (NL).

Breheny, M. (1996) 'Centrists, Decentrists and Compromisers: Views on the Future of Urban Form', in: M. Jenks, E. Burton and K. Williams, *The Compact City: A Sustainable Form?* E&FN Spon, London, pp. 13–35.

Davies, H.W.E. (1989) *Planning Control in Western Europe*, HMSO, London.

De Roo, G. (2000) 'Environmental planning and the compact city – a Dutch perspective', in: G. de Roo and D. Miller (eds), *Compact Cities and Sustainable Urban Development; A critical assessment of policies and plans from an international perspective*, Ashgate, Aldershot (UK), pp. 31–42.

De Roo, G. and D. Miller (eds) (1997) 'Transitions in Dutch environmental planning: new solutions for integrating spatial and environmental policies', *Environment and Planning B: Planning and Design*, Vol. 24, pp. 427–436.

De Roo, G. and M. Schwartz (eds.) (2001) *Omgevingsplanning, een innovatief proces; Over integratie, participatie, omgevingsplannen en de gebiedsgerichte aanpak*, Sdu Uitgevers, The Hague (NL).

Dryzek, J. (1990) *Discursive democracy; Politics, policy and political science*, Cambridge University Press, Cambridge (UK).

Etzioni, A. (1967) 'Mixed-scanning, a 'Third' approach to decision-making', in: *Public Administration Review*, Vol. 27, pp. 385–392.

Faludi, A. (1987) *A Decision-centred View of Environmental Planning*, Pergamon, Oxford.

French, N. (2002) *The Red Room*, Penguin Books, London.

Friedmann, J. (1987) *Planning in the Public Domain; From Knowledge to Action*, Princeton University Press, Princeton, New Jersey (US).

Friend, J.K. and N. Jessop (1969) *Local government and strategic choice*, Pergamon, Oxford.

Greiving, S. (2001) 'Afstemming en integratie tussen ruimtelijke en waterhuishoudkundige planning in Duitsland' [Coordination and integration between spatial planning and water management in Germany], in: G. de Roo en M. Schwartz (eds) *Omgevingsplanning, een innovatief proces; Over integratie, participatie, omgevingsplannen en de gebiedsgerichte aanpak* [Area oriented comprehensive planning, an innovative process; About integration, participation, comprehensive plans and the area-oriented approach], Sdu Uitgevers, The Hague (NL), pp. 217-230.

Healey, P. (1983) '"Rational method" as a mode of policy information and implementation in land-use policy', *Environment and Planning B, Planning and Design*, Vol. 23, pp. 217–234.

Healey, P. (1996) 'The communicative turn in planning theory and its implications for spatial strategy formation', *Environment and Planning B, Planning and Design*, Vol. 23, pp. 217–234.

Innes, J.E. (1995) 'Planning Theory's Emerging Paradigm: Communicative Action and Interactive Practice', *Journal of Planning Education and Research*, Vol. 14(3), pp. 183–189.

Jenks, M., E. Burton, K. Williams (eds) (1996) *The Compact City: A Sustainable Form?* E&FN Spon, London.

Lindblom, C.E. (1959) 'The science of muddling through', *Public Administrator Review*, 19, pp. 78–88.

March, J. and H. Simon (1958) *Organizations*, Wiley & Sons, New York.

Meyerson, M. and E. Banfield (1955) *Politics, Planning and the Public Interest; The case of public housing in Chicago*, Free Press, New York.

Rittel, H.W.J. and M.M. Webber (1973) 'Dilemmas in a General Theory of Planning', *Policy Sciences*, 4, pp. 155–169.

Sager, T. (1994) *Communicative planning theory*, Avebury, Aldershot (UK).

Sagoff, M. (1988) *The economy of the earth: Philosophy, law and the environment*, Cambridge University Press, Cambridge (UK).

Simon, H.A. (1960) *The New Science of Management Decision*, Harper & Row, New York.

Van den Broeck, J. (2001) 'Geïntegreerd beleid in Vlaanderen: Ideeën, voorstellen en lessen' [Integrated policy in Flaenders: Ideas, proposals and lessons], in: G. de Roo en M. Schwartz (eds) *Omgevingsplanning, een innovatief proces; Over integratie, participatie, omgevingsplannen en de gebiedsgerichte aanpak* [Area oriented comprehensive planning, an innovative process; About integration, participation, comprehensive plans and the area-oriented approach], Sdu Uitgevers, The Hague (NL), pp. 203–216.

Voogd, H. (1995) *Methodologie van ruimtelijke planning* [Methodology of spatial planning], Couthino, Bussum (NL).

Woltjer, J. (2000) *Consensus in Planning,* Ashgate Publishing Ltd, Aldershot (UK).

Chapter 7

Understanding Fuzziness in Planning

Gert de Roo[1]

7.1 Introduction

> Unfortunately, neither 'determinacy' nor stability are features of social reality. The more we reach for commonality in human interactions, the farther away it seems to be (J.F. Lyotard 1984).

In this chapter we argue that disappointments in contemporary planning are partially the result of taking for granted the understanding of crucial notions, because one is unaware of the fuzziness that surrounds them. We suggest that planning processes will improve substantially when this fuzziness is taken into account. In order to do so, the meaning and understanding of notions and concepts must receive greater attention. This implies that actors should be willing to work towards a mutual understanding of these notions and concepts, preferably at all the stages of the planning process. Three aspects are then to be addressed: fuzziness in planning that needs to be understood and needs to be challenged; actors within the planning arena, who have to come to mutual understandings about how they might work together, in order to cope with fuzziness; and – last but not the least – a mechanism, tool or method to identify and to tackle the fuzziness for individual processes in planning. Such a method will receive full attention in Chapter 8. In this Chapter 7 we will look at fuzziness as a phenomenon, how it can be understood and how actors can deal with it. In doing so, we will introduce actor consulting as a useful approach to cope with fuzziness in planning, and the principal position actor consulting might take within the realms of planning thought.

We start in Section 7.2 by building on the arguments of Chapter 6, stating that while the spectrum of planning thought is already covered fairly well, there is still a need for specific decision-making models. We argue that actor consulting – although in itself not a new approach – might be welcomed as a model, as it could fill one of the gaps that is apparent within the spectrum of planning thought. Section 7.3 explains that this particular model is needed to cope with the indistinct nature of the planning process and the role of the actors involved. Before going into the model itself (Chapter 8) we try to identify in Section 7.4 when and where fuzziness might be expected. We will conclude in Section 7.5 that fuzzy planning issues can be categorised as being either

1   Gert de Roo is Professor in Planning at the Department of Planning and Environment, Faculty of Spatial Planning at the University of Groningen, Groningen, The Netherlands.

'complicated' or 'complex'. This differentiates them from simple and straightforward issues and issues to be considered *very* complex or chaotic. Finally, in Section 7.6, we will attempt to grasp the meaning of fuzziness itself.

7.2   Towards actor-consulting

The argument in Chapter 6 makes clear that the view of planning as a technical exercise is far too limited. In contemporary planning theory it is widely accepted that content and interaction, and facts and values, are intertwined. As Forester (1989) puts it:

> It becomes clear that planning problems will be solved not solely by technical experts, but also by pooling expertise and non-professional contributions too; not just by formal procedure, but also by informal consultation and involvement; not predominantly by strict reliance on data bases, but also by careful use of trusted resources, contacts, and friends; not mainly through formally rational management procedures, but through internal and external politics and the development of a working consensus; not by solving an engineering equation, but by complementing technical performance with political sophistication, support-building, liaison work – all this, organizing – and, finally, intuition and luck (p.152).

In retrospect (Figure 7.1), we see two extreme approaches: 'technical-rationality' for very simple cases, with a high degree of certainty and control, and 'communicative-rationality' for very complex cases, where uncertainty prevails, particularly in relation to actors' perceptions, motivations and behaviours.

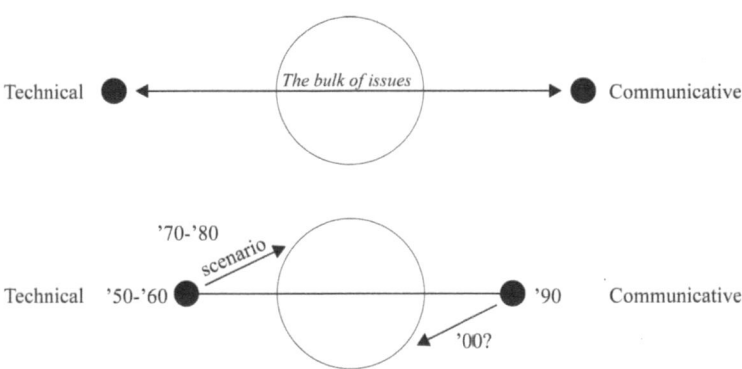

Figure 7.1   Development and progress in planning theory and practice (De Roo 2002)

Communicative interaction is important since it is necessary that governments deal with the strategic behaviour of actors and their means to influence decision-making processes (See e.g. Kaiser et al. 1995). In contemporary Western societies, governments have no full control. Instead, they are dependent on other actors to achieve their private and public aims. In practical terms, communicative interaction is necessary to facilitate an understanding of the aims, means and behaviour of other actors. This enables governments to anticipate and to incorporate the actions of others.

However the acknowledgement that there are two extremes in planning thought – technical and communicative rationality – is not enough (De Roo 1999, 2003, Van der Valk 1999). When focussing on what lies between the two extremes (the bulk of the issues in need of a planning strategy are found here), we will notice a change in balance, while moving from the technical towards the communicative extreme, between an object-oriented focus on content and goals, and an institutional-oriented focus on interaction (see Figures 7.1 and 7.2).

The technical rationality and communicative rationality approaches are both idealised types of planning, designed to handle issues rarely found in reality. These idealised types of planning however do not only represent strong belief systems: they are also important in understanding how planning theory relates to practice. Nearly all issues in planning practice include technical or object oriented aspects as well as communicative or institutional aspects. By far the majority of issues cannot be dealt with by either a purely technical or a communicative approach. Our point here is that although we have these idealised approaches at hand, their positions are rather extreme when compared to the circumstances of most planning issues. Therefore the challenge is to identify approaches to tackle these issues in between both extremes.

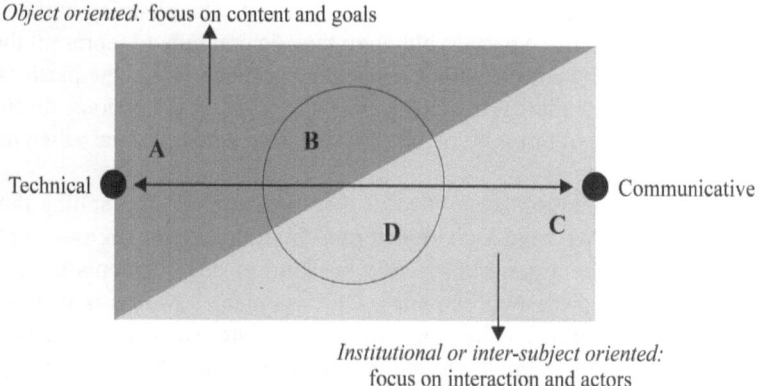

Figure 7.2   Orientation and focus of mainstream approaches in planning (for A, B, C and D see text)

Let us examine the characteristics of Figure 7.2. The points A and C represent theoretical planning situations, at the respective extremes of the technical and communicative spectrum of planning issues. The technical approach (A in Figure 7.2) takes the planning issue as predefined, and focuses on content and goals, resulting in blueprint planning. Certainty and control are present, due to the direct relationship between cause and effect, making it fairly easy for a single actor – normally the national government or the local authority – to decide how to plan and act. This approach works nicely in routine situations, however fails as soon as key information is not visible to the planners, or is withheld by one or more crucial actors. In these cases the technical approach proves to be nothing more than an ideal but unrealistic mode of planning.

The scenario approach (B in Figure 7.2) tries to respond to these weaknesses, although the issue is still taken as predefined, and the focus remains on content and goals. The difference lies in the acceptance that various routes in the planning process might follow. The issue might be predefined, but the outcome becomes situation and context dependent. The result is a cyclical process with evaluative feedback loops that keep the process on track to reach a desirable outcome. Instead of using blueprint plans, plans are often tactical progress reports. This planning model can be seen as a response to the lack of certainty in planning and therefore a response to a 'bounded' rationality, instead of an acceptance of uncertainty.

The rise of the communicative or participative approach in planning (C in Figure 7.2) is in a way an implicit acceptance of uncertainty. The focus of this approach is on interaction, to get consensus and commitment about the nature of the issues, how actions to tackle the issues can be taken, and by whom. The central idea here is that the interested parties should talk together about their mutual interests. These discussions might result in the production of an action plan, which is in effect a contract of commitment between the parties involved. The planning authorities are in a way equal among these parties, although they do continue to represent the wider collective interest. The premise of this planning approach is that the parties have to be brought together to address the uncertainties inherent in the actions, motivations, roles and perceptions of those involved. The resulting agreements are then used as a basis for further action.

With the above analysis it might seem that the spectrum of planning thought is covered fairly well. We have a good picture of both the ideal extremes in planning thought. We also have a good image of what kind of developments have evolved in the past between those two extremes. Our argument however is that while the communicative approach touches the institutional, inter-subjective and interactive dimension of planning – indeed a major step forward in planning – being rather extreme it should not be regarded as the *only* approach to all those issues that appear to require an institutional or interactional approach. As with the technical rational approach and the scenario approach that evolved from it, alternative approaches might be needed that build on the foundations of communicative rationale (as expressed in Figure 7.2). With the introduction of an actor-consulting model we consciously try to address this matter.

7.3 Actor-consulting as a means of coping with fuzziness in planning

An important argument here is that in practice we are still unable to deal with all planning issues and situations effectively. This is especially valid for concepts such as 'sustainable development', 'the compact city' and 'participation'. These fuzzy or fluid concepts tend to have in common the acceptance by the many, and consequently appear regularly in policy documents. Nevertheless, the outcome of the planning that takes place under the influence of these concepts often tends to be disappointing.

As argued in Chapter 1, it is most likely that many of our common notions, concepts and doctrines in planning are not understood as well as we implicitly like to think. Sustainability is one of the most obvious. While sustainability is embraced by society, it often remains unclear how it might be achieved in practice (O'Riordan and Voisey 1998, Jenks et al. 1996). The broad acceptance of sustainability makes it almost logical to follow a top-down, technical-rational approach. However, in doing so, the implementation of sustainability leaves substantial room for multiple interpretations. Rather slowly we are beginning to understand that sustainability is not so much a clear and given notion, as it is a declaration of intent that needs considerable discussion before implementation can proceed. Sustainability is consequently one of those notions we can consider to be fuzzy by character.

This fuzziness in planning is the result of the indistinct nature of planning objects and planning processes. Healey talks in Chapter 2 about the indistinctness of the actors' perception. Martens (Chapter 3) sees indistinctness as a result of a changing governance environment. Voogd and Woltjer discuss indistinctness in the role of decision-makers. In Chapter 5 Miller argues about the indistinctness of the objectiveness of information. As stressed throughout this book, there is also uncertainty involved in moving from strategic decisions towards implementation. A key issue here is the extent to which actors might conform and perform according to decisions made. All these uncertainties contribute to fuzziness, fluidity and illusiveness in planning.

The question arises as to what mechanisms lie behind this indistinctness. In other words, how can we describe or characterise the difficulties concerning generally accepted but fuzzy notions, concepts and visions in planning (such as 'sustainability'), which are disappointing in their performance? Apparently the discourse representing these notions and concepts still expects the ideal to match with the reality, whereas in fact it represents belief systems that fail to grasp reality due to hidden, uncontrollable and indistinct processes.

What in today's planning is missing – or not yet considered as relevant – are decision-making models, which are capable of handling issues where the content and goals are more or less accepted, but where an insight into the way actors should act, could act, or are willing to act is lacking. Despite the fact that in today's planning arena these issues are widespread, a common understanding of how to tackle them has not been achieved, so far... Therefore there is the need to review and to re-evaluate the individual as well as the common frames of reference. This opens the way for reintroducing the technique of actor consulting, which can be seen to occupy the position indicated by D in Figure 7.2. Clearly we are arguing here that while the technique itself is not new, a re-evaluation of its use in planning practice and its position in planning theory might be needed.

7.4   Fuzziness in planning: When and where?

Obviously, if we want to be convincing in arguing that actor consulting should be considered to be a useful method to tackle fuzziness, we have to be explicit about fuzziness as a phenomenon. To understand fuzziness in planning, we have to move away from our traditional mode of thinking about planning. We can no longer stay within the realm of modernistic, black or white, and 'yes' or 'no' thinking. In our deliberation we have accepted the thought that a technical rationality approach is an acceptable way forward under certain limited situations in planning. And in our quest for alternatives we have embraced the so-called communicative rationality approach as another extreme on a spectrum of planning rationalities. Technical approaches on the one hand are useful in the case of relatively straightforward issues, communicative approaches would suit the very complex (or chaotic) situations.

In other words, we are using complexity as a keyword to link these two approaches in planning, as it refers to the various uncertainties in planning. While the technical approaches are based on ideas that all crucial knowledge is or will be made available, and certainty is all around, the communicative approaches are based on the idea that the world of planning and decision-making is highly complex due to uncertainties in how the various actors will behave in a planning process. In communicative approaches each actor can be seen as a black box, making it unclear how each is going to respond. When dealing with a number of more or less independent actors, the black boxes become numerous, and the complexity of the issue increases rapidly. Instead of formulating targets, goals and objectives, an approach based upon consulting, commitment and consensus provides a more appropriate planning process. As such the *degree of complexity* (not to be confused with so-called 'complex systems') becomes a criterion for choosing the mode of planning and decision-making (see Byrne 2003, De Roo 1999, 2001, 2003, Christensen 1984).

How does this relate to fuzziness and – more importantly – to our understanding of what fuzziness is? By introducing complexity, are we not going towards an understanding of fuzziness that is too sophisticated? We do not think so. With regard to fuzziness, we have two motives to address complexity as a criterion for planning and decision-making. First of all, using complexity as a criterion for decision-making we can pinpoint a category[2] of planning issues that is likely to suffer most

---

2   As both Rosch (1978) and Lootsma (1997) state, a 'category is [...] a collection (set) of objects (items, elements) which are considered to be equivalent: one does not differentiate between objects in a category because differentiation is irrelevant for the purpose at hand. Categories do not have clear-cut boundaries. A category is usually defined, not by its boundary or by an exhaustive list of its objects, but by one or more highly prototypical (highly characteristic) objects' (Lootsma 1997: 13). This addresses the human observer rather well,

from fuzziness: the so-called 'complex' planning issues (again, not to be interpreted as the so-called 'complex systems' referred to in complexity theory). Our second motive is that complexity and fuzziness are very much related, with fuzziness being part of and contributing to complexity (See § 7.6). We will start by exploring our first motive: that of pinpointing the category of planning issues that is most likely to suffer from fuzziness.

So far we have understood fuzziness to be representative of a situation that is unclear about the precise meaning of notions and concepts, and the consequent response of actors. This is not a situation to be expected in a technical rational environment. It is very much a situation in which the actors would be likely to support an inter-subjective, institutional or communicative approach. A communicative rational approach seems within reach, as it is an approach that is most successful in situations that might be considered very complex, if not chaotic. Fuzziness is likely to be a prominent feature in this particular situation.

Paradoxically, we are willing to argue that in this situation, fuzziness is not something we should be bothered about. In situations requiring a communicative rational approach, many aspects (as well as the underlying notions and concepts) are likely to be unclear. Therefore such a situation calls for an open agenda at the beginning of the planning process. This open agenda is needed to work towards agreements about almost every aspect that is part of this process. In doing so, the intentions of each and everyone participating in the process will be addressed. With clear confirmation about the intentions of all, with a well defined frame of reference and with consensus about a common understanding and shared responsibilities, a common interpretation of notions and concepts is likely to be expected as an outcome of the communicative and participative process. Fuzziness is dealt with implicitly.

The communicative approach tends to be useful in very complex situations, due to the diversity and interests of the parties. In this situation, the various parties tend to have more or less the same, but opposing, interests. The communicative approach focuses on reaching consensus with respect to a common understanding. Before the parties can do anything together, an agreement on the precise nature of the issue is needed. Conformity is therefore the first step to be taken. In that sense we are talking about a first order consensus: actors agree or do not agree.

In contrast to this situation, in planning issues for which actor consulting is seen as a useful approach, the interests of parties are likely to be more or less in line with each other, even before the planning process has started. Consensus or agreement will already be at an advanced stage of development. In that sense, conformity concerning intentions has already been reached. As such actors easily agree that they agree, which is seen as a second order consensus. What we doubt however is that the actors have a full awareness about what precisely they might be agreeing upon!

---

as he 'reduces an infinite number of possible objects to a small number of collections of practically equivalent objects' (Lootsma 1997: 14). In other words, humans do understand the reality they are confronted with by generating categories to structure their observations and interpretations.

If actor consulting is seen as the way to go, a common agreement or vision can usually be used as a starting point. The main point will be that still much has to be done to substantiate this common agreement or vision. As indicated above, substantial knowledge is needed, not so much about intentions, but about what in practice might be done to achieve these intentions. The role, perception, motivation and behaviour of the parties involved at the various stages in the planning process are becoming important, with respect to the intentions all have (implicitly or explicitly) agreed upon.

Following this line of reasoning, we can address those planning issues that are fuzzy by character, and which will remain fuzzy if not addressed. These issues are likely to be found in between the categories that could be identified as rather simple and very complex. In between these categories the category of 'complex' planning issues remains. This is likely to be the category in which fuzziness might require explicit attention, by taking for example an actor-consulting approach.

We consider actor-consulting to be useful within the category of complex planning issues. Within this category the scenario approach has also been considered to be effective. The scenario and actor-consulting approaches have in common the implicit acceptance of at least a degree of uncertainty, in situations where certainty seems to be almost there for the taking. However there is a fundamental difference between the actor-consulting approach on the one hand and the scenario approach on the other. Using the scenario approach the actors involved are not likely to be consulted, irrespective of the nature and possible outcome of the scenario. The scenario approach is used when uncertainties are faced in relation to the *object* of the planning exercise (usually the physical environment). In contrast to this, the actor-consulting approach hardly questions the outcome of planning. There are however uncertainties in relation to the actors participating in the planning process. While the scenario approach brings the outcome of the planning process up for discussion, the actor-consulting approach questions the institutional framework.

Using categories of planning issues that vary in complexity we are beginning to get a hunch about the nature of fuzziness in planning. In this sense 'complexity' provides a criterion to understand when and where fuzziness might appear, when caution is needed, and when the consulting of actors (see Chapter 8) might be a wise approach. There is however more to it, and here we come to our second motive in addressing why complexity can help to understand fuzziness.

7.5   Fuzziness in complicated and complex situations

Within the category of complex planning issues, we made a distinction between classic, object related issues, in which we would use a scenario approach, and issues we believe are very much actor related. However, in our quest to understand fuzziness there is another distinction we would like to make. This again embraces the concept of complexity.

Having a closer look at this category of complex planning issues, we have to differentiate between the 'complicated' and the 'complex'. Let us firstly examine the nature of 'complicated' planning issues. From a traditional, reductionist point of view one might be tempted to model the reality of a planning issue by focusing on more and more detail, and by expanding the amount of indicators and relations between them (Emery and Trist 1965, Kramer and De Smit 1991, De Roo 1999, 2003). The argument here is that by breaking down a planning issue into its component parts and by focusing on the relations between the parts, a stronger foundation to cope with reality might be the result. In 'complicated' situations a pursuit of certainty does make sense, although one must bear in mind costs and time constraints. A situation can be described as 'complicated' when direct causal relationships are still demonstrable, only one has to dig substantially to overcome the obstacles.[3] These situations are not uncommon. Take for example a municipality that wishes to promote sustainable housing construction projects, without considering carefully the implications of this intention. Aspects of spatial and environmental policy to support the notion of 'sustainable housing' might for example include consideration of spatial location, recycling of land, transport to work, access to green spaces, local ecology, building materials, community development, energy conservation and efficiency, and so on (see Chapter 11). At a more detailed level, it might be necessary to consider if the use of tropical wood is considered acceptable (for its qualities as a long lasting, and as such sustainable material) or excluded (as it contributes to the decline of tropical rain forests, which is considered to be unsustainable). With 'complicated' planning issues, disappointing outcomes can be explained by the adoption of an over-simplistic view or by the failure to agree at a sufficiently detailed level the planning goals, instruments and actions to be taken.

The main point of this book is not to take the understanding of common notions and concepts too easily, in believing that intentionally all actors do agree upon these notions and concepts. Under complicated conditions, it is understandable that actors might not have a desire to dig too deeply into the nature of notions and concepts. Obviously, our point is that commonly understood notions and concept are not an automatic guarantee that actors will behave in support of each other's actions. On the contrary: without any consultation, fuzziness will become a major constraint in handling complicated issues.

Then there are *complex* planning situations. The difference between 'complicated' and 'complex' issues can be typified by – among other things – the appearance of remote causal relations (see § 7.6) and the fluid character of entities. As a consequence, in complex cases multiple interpretations are a reality and cannot be ignored. In these cases uncertainty is a fact. While under complicated conditions fuzziness might be excluded if actors work together, there is no escape under complex conditions.

---

3   Obstacles might include, for example, a lack of public understanding of the complexity of the possible effects of a development proposal. The case study at Chapter 9 describes how actor consulting was used to help to justify the use of an indicator of water quality as a means to develop public and political debate.

Under complex conditions fuzziness has to be considered a factor of importance and arrangements have to be made to deal with it. Here notions and concepts are indeed nothing more than declarations of intent, which need substantial consultation among actors to reach a common understanding and a common approach.

The 'compact city' concept, for example, promises car reduction, less energy consumption and other environmental benefits, whereas in practice we might find that a socio-demographic change – for example the rise of double-income families – causes intensification of traffic, leading to a growth in energy consumption, pollution and congestion. Again, sustainability is an obvious example due to multiple interpretations of its meaning, which are unsuitable to take forward towards implementation. The result is for example that actor A believes that cycling is contributing to a sustainable environment while actor B would prefer an improved highway as it would support a sustainable economy. Therefore agreements made at the beginning of a decision-making process are important to get commitment about how sustainability will be interpreted and how goals will be met. Instead of using objective indicators as criteria for how goals are going to be met, the result will depend largely on the actors involved and their respective understanding of the meaning of 'sustainability'. Clearly a shift from facts towards values and opinions must be contemplated.

Sustainability can be expressed only partially in objective terms, and even these can be heavily disputed. Are facts and interactions really what they appear to be? The question of global warming for example teaches us that in understanding the physical world (surely of all things an objective concept?) intuition, fear and belief count as much as facts and figures. Choices are made on the basis of both facts and arguments, which underlines the importance of a thorough and realistic decision-making process.

Almost all planning issues lie between the two extremes of technical and communicative rationality approaches. Transforming this statement into terms used by complexity theory (See § 7.6), we could say that almost all issues in planning are in between order and chaos, and are neither fully certain nor fully uncertain. All issues feature some element of uncertainty. In *complicated* situations more certainty might be created if efforts are made with substantive action. However uncertainty in *complex* cases cannot be ignored. While it might be worthwhile to pursue certainty, it is important for planners to find the right balance between certainty and uncertainty. Planners therefore will make a major step forward if they accept uncertainty as being a real and unavoidable part of complex planning processes.

Using the 'degree of complexity' as an indicator allows us to understand what fuzziness might mean. By making this differentiation between complicated and complex cases, we now are in the position to identify situations in planning where fuzziness might occur (complicated) and where fuzziness is unavoidable (complex).

7.6 Understanding fuzziness

We now have an understanding about when, where and how fuzziness in planning might appear. We still have not touched upon the nature of fuzziness itself. What is fuzziness and how should we understand its implications? To find an answer, again the concept of complexity will be used, and discussed in terms of a variety of theoretical foundations.

Why are we making such a fuss about this concept of complexity? So far we have debated the degree of complexity, which can be seen as a measure for uncertainty, consequently adding to the difficulty of predicting outcomes of actions and interactions. Another answer is that there is something more to complexity than its being a criterion for planning and decision-making. By this we mean the way complexity can be understood in the light of what is often referred to as 'complexity theory' (Byrne 1998, Eve et al. 1997, Kauffman 1995, Prigogine and Stengers 1984, Waldrop 1992). This theory is referring to so-called 'complex systems', representing the idea that nothing in the world is fully and permanently stable, and as such uncertainty is almost always around. Our material and social world is seen as a dynamic process, seeking the right balance of stability and fluidity (Waldrop 1992; 308). Complexity theory is therefore a form of conceptual thinking that embraces fluidity, fuzziness and uncertainty, and which attempts to cope with these aspects of reality.

Complexity as it is used in complexity theory is often understood in the light of causality. However, our discussion about fuzziness in planning brings another aspect to attention: complexity might not only be understood in terms of causality (or the lack of it) but in terms of fuzziness as well. This brings us to the second motive, expressed in Section 7.4, which involves using complexity to understand fuzziness. Complexity and fuzziness are very much related, with fuzziness being a part of and contributing to complexity.

In most literature, complexity, certainty and uncertainty are related solely to causality. Causality represents the logic of explaining how the cause of events takes place ($A \rightarrow B$). *Reductionists* and *positivists* believe that if you dig deep enough into the circumstances of a problem then eventually direct causal relations will be seen, which will lead to a certain and 'true' understanding about how reality 'works'. This modernistic perspective is being challenged here, as we have argued that uncertainty is part of reality as well. From this point of view uncertainty can be seen in terms of the impossibility of observing direct causality or direct causal relations (*causa proxima*). Instead, indirect or remote causal relations (*causa remota*) will be seen, resulting in unpredictable events. These unclear relations therefore contribute to uncertainty and add to the degree of complexity.

In this situation, an event or the occurrence of an event can be seen as representing a transition of one entity into another. The *event* is a term to express the 'becoming' while the *entity* stands for the 'being'. Apparently our society, in a continuous flux or state of change is focussed most of all upon events taking place, leaving the entity rather underexposed. The entity – and as such the 'being' – is however an issue that relates to complexity and uncertainty as well.

According to Aristotle's binary logic an entity can be described as 'A', as a clear opposite to 'not-A' (see Kosko 1993). In other words, in this line of reasoning 'A' can never be the same as or will not have any overlap with 'not-A'. This is also known as the 'law of non-contradiction (Lootsma 1997: 5). The point we are making here however is that this view is considered to be limited and unrealistic. Some appearances or entities are simply less clear than they might seem. Entities quite often appear to be fuzzy. Fuzzy means 'A' is not necessarily clearly defined, and could as well be in overlap with 'not-A'. The main concept behind fuzziness, in the words of Bandemer and Gottwald (1995) can be '... most easily grasped if one has in mind that in everyday life [...] one does not directly meet sets with a crisp 'borderline', but quite often it seems that there exists something like a gradual transition between membership and non-membership'.

There is multi-valence, or 'vagueness' as Bertrand Russell used to call it (Kosko 1993). And that is precisely what fuzziness is.[4] Kosko (1993: 19) states: 'The opposite of fuzziness is bivalence or two-valued-ness, two ways to answer each question, true or false, 1 or 0. [...] It means analogue instead of binary, infinite shades of grey between black and white. It means all that the trial lawyer or judge tries to rule out when saying, 'Answer just yes or no'.' This means that fuzziness contradicts concepts of a 'true or false' nature that exclude intermediate values (Lootsma 1997: 5). With this statement we are touching the academic debate of *fuzzy logic*.[5] As such, 'fuzzy uncertainty does not concern the occurrence of an event but the event itself, in the sense that it cannot be described unambiguously' (Munda 1995). *Fuzziness means multi-valence.*

---

4   In 1965 Lotfi Zadeh did put the label 'fuzzy' on vague or multi-valued sets known in mathematics (Zadeh 1965). Starting as a mathematical theory of vagueness, 'fuzzy set theory' has become a research area that has attracted mathematicians, economists, physicists and social scientists. Mathematicians have worked on models based on 'fuzzy sets', trying to "express the amount of ambiguity in human thinking and subjectivity [...] in a comparatively undistorted manner' (Terano, Asai and Sugeno 1992: 1). Within the field of social sciences decision-making is one of the major subjects of interest. See Kickert 1978 and Zimmermann 1987.

5   Fuzzy logic, fuzzy-set theory or fuzzy science originate from the work by Zadeh (1965), representing 'an attempt to construct a conceptual framework for the systemic treatment of vagueness and uncertainty both qualitatively and quantitatively' (Dimitrov 1997:1) and as such is – among others – about coping with social complexity. In essence fuzzy logic is about processing imprecise information. It is however criticized 'for being probability theory in disguise' (Lootsma 1997: 5), as it is trying to come to well-defined categories and to model degrees of truth. While true or false are considered to belong in a modernistic world, fuzziness is seen as a reality within a post-modern world. Obviously a mismatch is the result. Nevertheless the exercise to identify workable categories with regard to an imprecise, vague or uncertain reality has brought us concepts such as 'the degree of concordance and discordance with certain assertions, and the degree of credibility of certain relations' (Lootsma 1997: 8).

According to Munda, Nijkamp and Rietveld (1992: 4) '... spatial systems in particular, are complex systems characterized by subjectivity, incompleteness and imprecision'. If we look back to Chapter 1 and the discussion regarding the difficulties encountered with the definitions of 'urban' and 'rural', and 'sustainability' and 'liveability', it is rather obvious that planning is full of fuzzy, multivalent notions and concepts. This explains the need to reframe our terms of reference regarding the notions present within a planning arena. This leads to the need to consult the various actors taking part in processes of planning that are not of an entirely simplistic nature.

Recapitulating, fuzziness in planning is likely to occur where complicated and complex situations meet.[6] On the one hand there is the possibility to achieve an element of certainty by investing in substantive action. In the case of complex situations, on the other hand, there is a rising degree of uncertainty, as a result of lacking causality, the existence of fuzziness or both. When fuzziness is around, the reframing of the perceptions of the actors involved should clearly precede any attempt to reshape our physical environment via planning goals and activities (De Roo 1999, 2003).

We understand at this point what fuzziness is, what it is caused by, where and when it can be expected and what the consequences might be. A tool is now needed to handle fuzziness. We will go on to argue that a process of consultation with actors is an important starting point in situations that are likely to suffer from fuzziness.

References

Bandemer, H. and S. Gottwald (1995) *Fuzzy Sets, Fuzzy Logic, Fuzzy Methods with Applications*, John Wiley & Sons, Chichester (UK).

Byrne, D. (1998) *Complexity Theory and the Social Sciences; An Introduction*, Routledge, London.

Byrne, D. (2003) 'Complexity Theory and Planning Theory: A Necessary Encounter', *Planning Theory*, Vol 2(3), pp. 171–178.

De Roo, G. (1999) *Planning per se, planning per saldo* [Planning per se, planning bottom up], Sdu Uitgevers, The Hague.

De Roo, G. (2001) 'Complexity as a Criterion for Decision-making; a Theoretical Perspective for Complex (Urban) Conflicts', paper presented at the 1st World Planning School Congress at the College of Architecture & Urban Planning, Tongji University, July 2001, Shanghai, China.

De Roo, G. (2002) 'In weelde gevangen: van ruimtelijk paradijs naar een leefomgeving in voortdurende staat van verandering...', [Caught in heaven: from a spatial paradise towards a state of continues change] Oration, Faculty of Spatial Sciences, University of Groningen, Groningen (NL).

---

6   In Chapter 5 Miller points to pragmatism, a philosophy that may help to cope with fuzziness in planning. In essence, pragmatism requires that we look at the consequences of an idea or decision.

De Roo, G. (2003) *Too good to be true: Environmental planning in the Netherlands*, Ashgate, Aldershot, UK.

Dimitrov, V. (1997) *Use of Fuzzy Logic when Dealing with Social Complexity*, Complexity International, Vol. 4, pp. 1–10.

Emery, F.E. and E.L. Trist (1965) 'The causal texture of organizational environments', in: F.E. Emery (ed.) (1969) *Systems Thinking*, Penguin Books, Harmondsworth, UK, pp. 241–257.

Eve, R.A., S. Horsfall and E.M. Lee (1997) *Chaos, Complexity and Sociology; Myths, Models and Theories*, Sage Publications, Thousand Oaks, US.

Kickert, W.J.M. (1978) *Fuzzy theories on decision-making*, Martinus Nijhoff Social Sciences Division, Leiden, Boston, London.

Forester, J. (1989) *Planning in the face of power*, University of California Press, Berkeley, USA.

Jenks, M., E. Burton, K. Williams (1996) *The Compact City: A Sustainable Form?*, E&FN Spon, London.

Kaiser, E.J., D.R. Godschalk and F.S. Chapin (1995) *Urban Land Use Planning*, University of Illinois Press, Chicago.

Kauffman, S. (1995) *At Home in the Universe; The Search for Laws of Complexity*, Penguin Books Ltd., London.

Kosko, B. (1993) *Fuzzy Thinking, the New Science of Fuzzy Logic*, Hyperion, New York.

Kramer, N.J.T.A and J. De Smit (1991) *Systeemdenken* [System thinking], Stenfert Kroese, Leiden (NL).

Lootsma, F.A. (1997) 'Fuzzy Logic for Planning and Decision Making', *Applied Optimization Series nr. 8*, Kluwer Academic Publishers, Dordrecht (NL).

Lyotard, J.F. (1984) *The Postmodern Conditions: A Report of Knowledge*, Manchester University Press, Manchester.

Munda, G. (1995) *Multicriteria Evaluation in a Fuzzy Environment; Theory and Applications in Ecological Economics*, Physica-Verlag, Springer-Verlag Company, Heidelberg (G).

Munda, G., P. Nijkamp and P. Rietveld (1992) 'Fuzzy Multigroup Conflict Resolution for Environmental Management', *Serie Research Memoranda*, Vol. 67, Vrije Universiteit, Amsterdam.

O'Riordan, T. and H. Voisey (eds) (1998) *The Transition to Sustainability: The Politics of Agenda 21 in Europe*, Earthscan Publications Ltd., London.

Prigogine, I. and I. Stengers (1984) *Order out of chaos*, New Science Press, Boulder, US.

Rosch, E. (1978) 'Principles of Categorisation': in E. Rosch and B. Lloyds (eds) *Cognition and Categorisation*, Lawrence Erlbaum, Hillsdale NJ (US), pp. 27–48.

Terano, T, K. Asai and M. Sugeno (1992) *Fuzzy systems theory and its applications*, Academic Press, San Diego (US).

Van der Valk, A. (1999) *Willens en wetens: planning en wetenschap tussen wens en werkelijkheid*, Wageningen Universiteit, Wageningen, NL.

Waldrop, M.M. (1993) *Complexity: the emerging science at the edge of order and chaos*, Viking, Penguin Group, London.

Zadeh, L. (1965) 'Fuzzy Sets, Information and Control', Vol. 8, pp. 338–353.

Zimmermann, H.J. (1987) *Fuzzy sets, decision-making and expert systems*, Kluwer-Nijhoff Publishing, Boston (US).

Chapter 8

Actor Consulting: A Model to Handle Fuzziness in Planning

Gert de Roo[1]

8.1 Introduction

> Civil knowledge is of all subjects the one that is most immersed in substance, and least reduced to axiom (Francis Bacon 1561-1626).

These words from the seventeenth century are still relevant to contemporary society. Civil knowledge provides a deep reservoir of meaningful information, based on citizens' experience and knowledge of their locale. However, much of this knowledge remains unexploited, or as Bacon says 'least reduced to axiom'.

Actors have their own opinions, value judgements, prejudices and assumptions, and these should be taken seriously within the planning arena. Their differences however add considerably to the fuzziness in planning. If we ignore the 'black box' that represents each of the actors involved, disappointments about the planning process and the outcome of planning are likely to result.

In this chapter we will introduce a planning tool that will help us to cope with actor-related fuzziness in planning. This tool will focus on the role, motivation, perception and behaviour of the actor within the planning process. Central to the tool is the consultation of the actor regarding his or her desired contribution, and his or her present or actual contribution towards solving a particular planning issue. This information will be supplemented by developing ideas for potential contributions to solve the planning issue. It will put the role of each actor into perspective, relative to the other actors involved, and the planning objective itself. The outcome of this actor-consulting process is meant to lead to a better understanding of the role, motivation, perception and behaviour of the actors. It will give a means to adjust the goals we set, and the regulatory instruments and the type of decisions we are able to use. It will also give us a platform on which we can criticise current planning processes, and if necessary, the institutional setting itself. The introduction of this actor-consulting model will present an effective way to bring actor's attitudes, motives and perceptions inside our range of vision.

---

1   Gert de Roo is Professor in Planning at the Department of Planning and Environment, Faculty of Spatial Planning at the University of Groningen, Groningen, The Netherlands.

This chapter builds upon Chapters 6 and 7. In Chapter 6 the theoretical context of our argument on fuzziness in planning has been elaborated. A spectrum of planning thought has been presented on the basis of which we were able in Chapter 7 to locate the occurrence of fuzziness in planning. We have identified when and where to expect fuzziness and how it is likely to appear. In Chapter 8 we will present actor consulting as a means to handle fuzziness in planning.

Section 8.2 locates the roots of actor consulting within the wider range of methods that relate to decision-making within a multiple actor arena. Section 8.3 links actor-consulting with contemporary sociological theory, on the basis of which – in Section 8.4 – coherent steps are constructed for consulting the various actors. This leads us to the composition of the model itself: Section 8.5 focuses in particular on desired, present and potential contributions of actors to notions, concepts, goals, targets and visions in planning. This approach is intended to lead to a collective agreement, which should improve the effectiveness of actions taken at the various stages in the planning process. In Section 8.6 it is argued that in order to maintain this collective agreement, there is a need for actor consulting to be carried out at *all* stages of the planning process. These arguments aim to justify the construction of the model and its potential to address the fuzzy character of notions and concepts in planning, The model will address the subjective nature of planning issues, develop a common understanding among actors, unravel underlying mechanisms that determine actions of actors, and in doing so will provide planning authorities with a proper means to anticipate uncertainties within the policy arena.

8.2   The origins of the actor-consulting model

Actor consulting is not a new approach, on the contrary. It is a fairly common approach both in public (governmental) and private (business management) planning practice. We however suggest that its links to the spectrum of planning thought as presented in Chapter 6, and its ability to address fuzziness in planning as explained in Chapter 7, makes actor-consulting an interesting platform for our argument. This has resulted in the development of a specific analysis method that addresses fuzziness in particular, based on the actual, the desired and the potential contribution of actors involved (see § 8.5).

As a method, actor consulting can be qualified in various ways. It is seen as a design approach for decision-making processes, it is seen as a problem structuring method for multiple stakeholder evaluation, and it is seen as a mediation technique for conflict resolution. With regard to the latter point, consultation is not considered here a means to cope with situations that are characterized by an unequal distribution of decision-making powers among actors. Consultation is often used in theories relating to conflict resolution, but in that sense is carried out by the decision maker (in case of public policy and planning this is likely to be an authority) to allow others affected by proposed (policy) measures to bring forward their opinions, while it remains up to the decision-maker only to incorporate the information gained into the final decision (Hanf and Koppen 1993).

In this book consultation is intended as a mechanism to work towards a well-defined mutual understanding and a common frame of reference between actors who are (implicitly) agreeing about the conditions under which the planning process will take place. As explained in Chapter 7, mutual adjustments are however needed due to the fuzzy character of (some aspects of) the planning process.

As such actor-consulting can be considered part of a family of problem structuring methods. According to Rosenhead (2003: 0, 2005) this group of methods 'provides support to diverse groups confronted with agreeing and making progress with a common problem'. It is a group of methods that provide model-based decision support for various problems that are somehow ill-structured. These ill-structured problems often include multiple actors, multiple perspectives, conflicting interests, and key uncertainties (Rosenhead 1996, Rosenhead and Mingers 2001). Rittel and Webber (1973) describe these as 'wicked' problems that beg for communicative approaches, in contrast to traditional and 'tame' problems that are more inclined towards technical solutions. The number of approaches is growing fast, and incorporates a wide range of possibilities, including methods such as Robustness Analysis (Rosenhead 1996), Drama Theory (Rosenhead 1996), Decision Conferencing (Watson and Buede 1987), Interactive Planning (Ackoff 1979) and System Dynamics (Lane 2000). These methods are considered to be of relevance to the field of planning, as they incorporate the 'engagement of human judgement and social interaction […] needed to make progress with problems for which […] simplifying assumptions are invalid' (Rosenhead 2003: 5). Actor consulting clearly belongs to this group of problem structuring methods.

Coming out of organization theory, actor consulting focuses on implementation and the 'how to do' question. Traditionally a phase model is used to mediate step by step towards recommendations for achieving commitment and for achieving better results. A model can provide a rational basis for this process, resulting in a plan that prescribes what activities should be carried out, and in what order. As we have considered in Chapter 6, this can be seen as a reductionist and a rather limited view, as it is trying to make rational what is likely to be uncertain and ambiguous.

However, actor consulting and other design methodologies have moved forward by incorporating a post-modern perspective, believing that when facing ill-structured problems, actors do not necessarily behave rationally but are coping by being reasonable. Therefore actor-consulting is focusing nowadays more on understanding and reframing the frames of reference of actors involved, instead of prescribing how to turn decisions into actions. No longer a predefined outcome (fact-finding) is being emphasized, but the use of actor consulting has shifted attention towards the *conditions* under which the planning process could take place. As such it is facilitating the planning process through comprehensive consultations that should result in deliberations between the actors about discrepancies and contradictions in perception and behaviour (Hanf and Koppen 1993). Actor consulting in that sense will lead to clarification of vague, fluid or fuzzy conditions at the various stages of the planning process.

Through actor consulting we will be able to structure a reality out of social processes, which are almost always chaotic by character. Outcomes of these processes 'are emergent, and not the direct result of the vision, plans and implementation efforts' (Visscher 2001: 23). Social processes, and more specifically governmental actions and their effects within a social and physical environment, are often lacking a predefined structure that incorporates a full understanding of what is going on and what is likely to happen, as they will almost never take place as isolated events. Processes and actions take place within a stream of ongoing events, and might interact and interfere with these events. And vice versa, processes and actions initiated will be influenced and structured by an almost invisible pattern of unwritten social laws, rules, constructions and processes, which we sometimes define as culture, and upon which we base our behaviour. To get an understanding of this invisible ground pattern and how it might structure decisions taken and the processes and actions that follow, a structure, a framework or a set of conditions that represents reality has to be composed.

We believe consulting actors about the stories they are willing to share can do this. It is felt that 'a structure is only created, before, during or after the action, when people tell stories about what will happen, happens or has happened' (Visscher 2001: 23, see also Harré 1975, MacIntyre 1980, Boje 1995, Cilliers 1998). Visscher explains, 'in a story, an unstructured stream of events receives a beginning and an end. Certain persons are singled out as main characters, separate actions are made visible, and actions receive meaning in light of a plot. Causes are attributed to circumstances and people with intentions. A story tells whether something happens, what happens, who makes a difference and why. From a muddled ball of events and actions, storytellers pull threads and weave them into a meaningful whole' (2001: 23). In other words, stories are what actors have to share and stories should be listened to, if we want to understand the motives, perceptions, behaviour and roles the actors are taking within processes of planning. In what Schön (1987) sees as a process of reflection-in-action, actors will be consulted for their stories, which enables us to design a structure or a framework upon which we begin to understand the conditions and assumptions under which decisions are going to be implemented.

Obviously the main purpose of this book is to emphasise the importance about how this knowledge will reflect upon the decision itself and upon what it stands for. Using actor consulting under the conditions explained in Chapter 6 and 7, we believe we might come to a commonly understood reality.

Central to reaching mutual understandings and common frames of reference is the importance of listening and of being heard (Stein 1994). Listening and being heard are two sides of the same coin that can be seen as a basis for change in both individual actors and the organizations they represent. The process of listening and of being heard reveals the storyline of the actor. This is common practice in a doctor – patient relationship. Ogden (1989: 16) writes that: 'the analyst has no means of understanding the patient except through his or her own emotionally coloured perceptions of, and responses to, the patient. Of these perceptions and responses, only a small proportion are conscious, and it is therefore imperative that the analyst learn to detect, read, and make use of his own shifting unconscious state as it unfolds in the analytic discourse'. This means that the one who is listening must have more than an open mind, but has to have a gut feeling about what is called 'inter-subjective resonance of unconscious processes' (Stein 1994). The importance of these conditions of listening and being heard is far reaching and includes the practice of planning.

8.3   Linking actor-consulting to sociological theory

The arguments above are closely related to the ongoing discussions within the field of sociological theory. An examination of sociological theory firstly helps to substantiate the basis of 'actor-consulting' and secondly helps to define the direction of the questions to be asked while consulting the various actors. For example, sociological theory tells us that: 'humans do not […] react in an automatic or mechanistic way to 'stimuli' or to objective circumstances, but must enter into a process of 'definition' or 'interpretation'' (Scott 1995: 101).

According to Blumer (1969: 5) 'the actor selects, checks, suspends, regroups and transforms meanings in the light of the situation in which he is placed and the direction of his action'. As such, the individual actor will always construct a meaningful definition of what he or she is focussing on, in close relation to its context. This context includes both the physical and the social environment. The contextual situation of our social environment helps us to colour our perceptions, and this in turn colours our knowledge. Obviously, this means that interaction with other parties is crucial to an understanding that will lead to relevant action. According to Scott (1995: 199) 'people do not 'know' which norms are to apply in a particular situation unless they have arrived at an understanding of the kind of situation that they are involved in. The interpretative procedures that people use allow them to build up a sense of social structure in which they are operating and, therefore, to identify the nature of the situation and the relevant norms that are to be employed'.

According to Luhmann (1982) there is a structural differentiation of three levels of relations and actions: the interactional, the organisational and the societal levels. He argues these have become relatively autonomous systems of action, each being subject to distinct mechanisms and processes. Luhmann argues that many of today's societal problems arise from dislocations between these levels. We can conclude that while the notions, concepts, goals and visions in planning can be held to represent society's needs and desires, it is the *implementation* of these that is failing, and it will be Luhmann's interactional and organizational levels that do not function properly. This means that 'actor-consulting' should perhaps focus on these two levels in particular and the way these interplay. The interplay will no doubt focus on the actors' *formal* and *informal*[2] roles within a collective, to reach consensus

---

2   See the UK case study at Chapter 11 ('Research Methods'), where it was necessary for reasons of political sensitivity to record only the formal views of the actors. Informal views,

about actions to be taken. Giddens (1984) adds to this by arguing that actors occupy 'positions' within institutions or groupings that make up their society, and that any analysis of interaction must recognise this relative positioning of the actors. Roles and identities that reflect these 'positions' contribute to a sense of social structure. The reverse is also true. People are able to account for what they are doing, if it is in line with the prevailing social structure.

This argument tells us that it is important to understand the social structure of parties involved, how these parties are positioned relative to one another, and the role each party is playing or is willing to play, both officially and unofficially. Understanding this social network is important to reach a common agreement, and to turn this agreement into fruitful and coherent forms of action. Scheff (1967), however, adds a few crucial remarks. He argues that what is needed, is what he calls 'reciprocated understandings' about a collective agreement. This could mean an infinite series of consultations. In such a situation, each participant knows that others agree, knows that they know that they agree, and so on. With this statement we have – again – reached an important step. While we believe we have gained mutual understanding about an issue, and how it has to be implemented and turned into action, this might still be not enough for a successful conclusion. While going through the various stages of a planning process, we might have to seek over and over again mutual understandings – conformity –, which at the end might lead to the conclusion that an acceptable performance has been reached. An ongoing dialogue is needed.

8.4   Steps to be taken

Here contemporary sociological theory meets our discussion about planning, process and performance. This shows us the possible steps to be taken to reach a meaningful and 'sustainable' conformance in planning. The following procedure can be distilled from the above discussion:

1. Identify the *contextual* structure in relation to the planning issue at hand. This structure is likely to focus upon the institutional, the cultural and the physical context of the issue. It will help to determine the context of the issue as a whole, and its constituent parts. At this stage it is important to define the planning issue on the basis of only those aspects that are considered to be essential, and that appear to be unclear or fuzzy.
2. Select the individual parties and organizations to be involved (the institutional structure). It is important to identify a limited number of crucial actors, who can be seen as representative of a wider population, and who are able to

---

however, were instrumental in understanding actors' motivations, and helped to frame and contextualise the actor contributions to sustainability.

contribute to a broader understanding of the issue.

3. Identify the set of conditions, such as definitions and interpretations of the issue at hand, the notions and concepts that are central to the issue and the policy system that is currently imposed upon it, by exploring the opinions of each of the individual parties and organizations.

4. Develop an understanding of the interactions and 'positions' of the actors within the planning arena. Identify the structures of meaningful relations between the interactional face-to-face level and the organizational level (identification of inconsistencies).

5. Identify the quality of these meaningful relations, in terms of whether it is likely that agreement might be reached regarding the definition of the problem, regarding a commonly accepted action programme, and regarding the division of tasks and responsibilities among the parties involved. If agreement is not clear then the issue may be more complex than expected. In this case a communicative approach may be needed instead of actor consulting. See Figure 7.2.

6. Repeat this process at each stage of or at every stage considered relevant to the planning process.

The analyst should be aware of the conditions that will frame the issue at hand and the institutional context that surrounds it. Kast and Rosenzweig (1974) consider several *subsystems* that are likely to emerge under the following five titles: 1. the technical subsystem; 2. the structural subsystem; 3. the psychosocial subsystem; 4. the managerial subsystem; and 5. the cultural and values subsystem. In consulting actors, an analyst will encounter stories that present a mixture of the various subsystems. In complex situations 'there never is just 'a problem'; there is always a complex problematique. Each individual is involved with his whole history, emotions and experiences. And, accordingly, each individual has a different perspective on the problematique' (in Visscher 2001: 93, quoting Yates). To structure, design or compose a well defined agreement or when constructing a new interpretation of an issue, an understanding of these subsystems might help to bring some order to the subjective stories (Ziegenfuss 2002).

We can offer some words of comfort with regard to the seemingly onerous nature of this process. For example, while the planning issue to be addressed might be seen as either complicated or complex, our theoretical platform has taught us that in an actor-consulting scenario the parties involved are most likely to present themselves in a transparent way. There is agreement to agree (second order consensus) and it is felt among actors there is agreement on the issue itself (first order consensus), although – as we argue throughout this book – due to good feelings about having a second order consensus the consequences of a first order consensus remains to be seen and is therefore fuzzy in character. Furthermore, since the concepts we have talked about so far – sustainability, compact city, and such – are likely to lead to initiatives within policy institutions, there is already substantial knowledge about the roles that the representatives of these policy institutions are likely to play. There are few hidden agendas to be expected. The willingness for conformance also leads to a relatively easy identification of the various actors and the formal 'positions' they adopt. This situation should not make it too difficult to come to a diagnosis of how the various actors consider their role in relation to the policy field under examination. However, this is also a situation in which prejudice so easily prevails, and in which it is easy to believe that actor consulting is not needed at all.

When the parties related to the planning issue at hand have been identified, the process of identifying the structures and the quality of meaningful relations can begin. However, our sociological theory tells us it is not at all clear how to reach a mutual understanding, on the basis of which actors can act as a collective. Here actor-consulting can be seen as a reflective and interactive method, that makes us not only conscious of the behaviour of the various parties, but also pinpoints the reason why they behave as they do ('reflection-in-action', see Visscher 2001). The main issue therefore is to identify a mechanism to extract this information during the actor-consultation process.

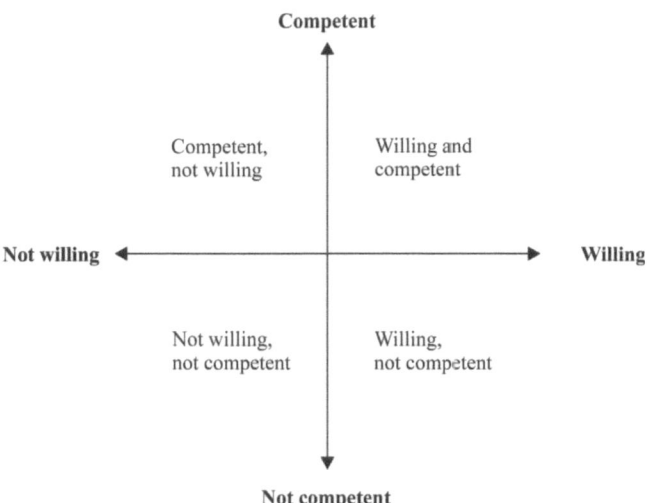

Figure 8.1   Willingness and competence as indicators for successful policy implementation

Source: CEA 1998

This brings us to the following approach. Earlier on we expressed the desire not only to answer the 'who' question but also the 'why and wherefore' questions (see § 6.2). An interesting suggestion is made by CEA (1998) to analyse the role and the likely behaviour of actors within a planning arena on the basis of two criteria: willingness and competence (see Figure 8.1). The background of this particular model was to see if authorities were ready to take up new responsibilities in a process of decentralization, in particular for dealing with soil contamination. Interestingly, quite a few actors were found in practice to be willing to take certain responsibilities (usually in the interests of power and money), however their competence to act was obviously lacking. Similarly, there were some authorities that were notably capable, but had no intention of taking over the responsibilities in case it would divert their attention from issues they considered more important.

Willingness and competence as indicators are interesting to explore. However, this technique needs an independent outsider to carry out the analysis and to select the criteria on which the analysis takes place. We are aiming to identify a tool that allows the parties involved to be reflexive towards their own attitudes and actions. According to Giddens, reflexivity 'involves the monitoring of actions – one's own actions and those of others – and the wider context within which these actions occur [as a means] to reconstruct the 'reasons' that lead to the actions.' Similarly, Scott sees actors as being continuously involved in 'the interpretative process of accounting for action and the normative process of being accountable' (Scott 1995: 205). This is exactly what we wish to achieve with the actor-consulting model, and in doing so we differentiate action not by willingness and competence but by the desired, actual and potential contributions of the actors.

8.5 Actors' present, desired and potential contributions

The technique of willingness and competence however does bring us near to our proposal. 'Willingness', in the sense of 'what an actor is willing to do' brings us to the '*desired contribution*' of a party. Asking for the desired contribution of an actor solves the difficulty of the analyst working with predefined criteria. Now it is the actor who can present his or her own criteria, as the desired contribution to, say, sustainability or the compact city policy. 'Competence' can be seen as how the various actors are able to contribute in terms of their '*potential contribution*' to planning.

This omits one important type of contribution. This is the actual or *present contribution* to planning. This provides an indication of the baseline conditions from the perspective of each actor whose history of experience must be confronted. This facilitates an understanding of the prejudices, attitudes and belief systems of each actor towards both the planning issue itself, and the role and behaviour of other parties. A lifetime of experiences, which are recognized within the context of a present contribution to a particular planning issue, can also be an important basis for complaints and frustration. 'Individual actors are able to modify their current actions in the light of results of their past actions', according to Scott (1995: 205). The present contribution can be the most important point of reflection, because an actor can contribute knowledge about the various barriers he or she is confronted with. Therefore the 'present contribution' allows the various parties to reflect upon what they are doing, and upon what they expect others to contribute.

Steps three to five (§ 8.4) allow us to reflect upon the various actors, their desires and their expectations. To obtain this information, it is important to know what actors are willing to do to contribute to the issue at hand (the desired contribution), their actual contribution to the issue (the present contribution), and how the different actors might be able in principle to contribute to the issue (the potential contribution). This model helps us to understand each actor's 'behaviour', from an interactional and an organizational perspective. It also teaches us how to treat an issue realistically, on the basis of the attitudes of the parties involved.

Actions taken by actors do not come about independently. They are determined among other factors by the resources available to each actor and the institutional setting in which they operate. These are dynamic because physical conditions change continuously, because interactions between actors influence changes of attitude, and because of the continuous development of institutional arrangements. This is the context in which actors will act in a particular way – their *present contribution*. Meanwhile they have certain ideas about the way they want to act – the *desired contribution*. The proposed actor-consulting model enables analysis of the present and desired contribution and the conflicts that might arise between them.

Additional literature study, expert meetings, introducing observation techniques and further analyses can generate information about the potential contributions of actors, and about the internal and external conditions under which such contributions are possible. Technical research might be of value here, such as the comparative study of a similar situation in other regions. Interviewing actors about their perception with regard to the role of other parties can add value to the research. This facilitates an understanding of the interdependence of actors that is needed to achieve objectives. Though interviewing actors about their perception of others can sometimes lack objectivity, it might generate ideas that are worthwhile pursuing. In short, we can distinguish two steps to study *the potential contribution*. The first step consists of finding out *what solutions* exist, while the second step refers to exploring the *advantages and disadvantages* of these solutions for different actors. This is akin to the exploration of *alternatives* that forms the basis of many rational planning techniques, such as that advocated by the Strategic Environmental Assessment Directive (European Commission, 2001: Article 5(1)).

This gives us the basic model, consisting of the *present*, *desired*, and *potential contribution* of actors (see Figure 8.2).

This approach is aimed at helping a planning authority to formulate well-considered, 'realistic' policy by reducing uncertainty. With the notion of 'realistic' policy we are at an important point of elaborating the actor-consulting model. Our approach must therefore reflect carefully upon the regulatory environment of the policy sector under investigation, and the relationship between the regulatory climate and the attitudes of the actors. We can describe this process of interaction between authorities and other parties in terms of *direct regulation, indirect regulation* and *self-regulation*. Direct regulation is usually imposed by a planning authority with legal powers, and may be implemented at national, regional or local levels. A local authority bylaw is a possible example. Indirect regulation aims to change behaviour

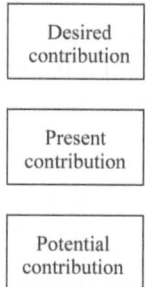

Figure 8.2   The basis of the actor-consulting model; the desired, present and potential contributions

by means of incentives, and usually takes the form of some fiscal instrument, or by the issue of subsidies or grants when certain criteria are met. Self-regulation generally consists of an agreement between actors to behave in a particular way.

Here we come to the essence of Luhmann's argument (1982) that many of today's societal problems arise from dislocations between what he calls the interactional face-to-face level and the organizational level. The planning authorities' desire to regulate our environment introduces frictions that we find difficult to understand due to the fuzziness of the issue at hand. If we can understand and overcome the fuzzy character of these issues, we will find ourselves in a position to propose more realistic regulation.

Martens in Chapter 3 points out that the classical coordinative model of governance is currently under pressure, and is being superseded by models that integrate elements of the competitive and communicative models of governance. This pluralistic form of governance will influence the mix of regulatory instruments that is likely to be effective. We could even say that this mixture is likely to become more pronounced as the mode of governance becomes increasingly fuzzy (see Snellen 1987: 18). It is important here to recognise the various forms of regulation, and to use them appropriately. Even more important is to compare the prevailing regulatory framework with the outcome of the analysis of desired, present and potential contributions. This will help to show us the structure and quality of meaningful relations.

The example in Chapter 14 relates to how the Province of Drenthe in the Netherlands, has a desire to 'become sustainable' within a period of 20 years. Whilst the provincial authority might in this example be tempted to use direct-regulation to pursue its objectives, an analysis via a process of actor consulting might show substantial objections to such an approach. It could even be that most parties are very willing to contribute to sustainable development (which is indeed the case in Drenthe), but do not want to act independently from other parties. This outcome could mean that a position between indirect regulation (for example by the issue of subsidies by the Province) and self-regulation (for example by means of a joint action plan between municipalities, site developers and housing organisations) might be far more effective in terms of performance than direct regulation. In the analysis to be conducted, we need to identify how the national government, regional authorities and municipalities could apply these three types of regulation successfully. Furthermore, we need to know how the actors perceive these regulations with regard to the objectives set for a certain issue.

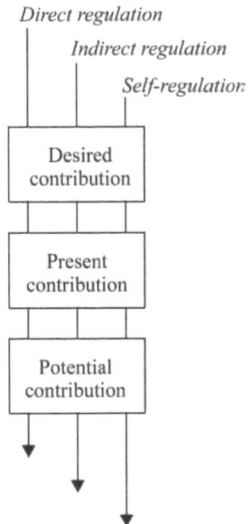

Figure 8.3   The actor-consulting model, including the different types of regulation

Figure 8.3 includes these three types of regulation. The arrows in the figure indicate that they should be considered at each step in the analysis. An important actor is the regulating authority, which wants to fine-tune its objectives and to identify the most effective set of regulations. Goals, targets and objectives, often based on fuzzy notions, concepts and visions, will also be included in their plans. The goals and objectives of the decision-making body play an important role, since these will be translated into formal policy by means of regulatory tools. Therefore it is not unlikely that the actor-consulting analysis is referring to the vision of the decision-making body itself. Obviously, if relevant plans from other organisations are available these should be included in the analysis. Subsequently, the outcome of the analysis will serve to indicate to the authorities the potential objectives they can pursue, and the types of regulation that support these objectives. Figure 8.4 depicts the completed model.

This model introduces the specific relevance of 'organizations', 'plans' and 'actors' within a particular planning context. These substantive criteria in particular are likely to clarify the storylines and the institutional context of the interests that are put on the table. These various interests have to be balanced, and together they will contributetowards an understanding of the entire issue. However, the interests need to be understood through the lens of the wider institutional framework. Consequently, the institutional background of organizations, represented by actors and presented through plans, must be analysed in order to gain a clear understanding of the issue at hand.

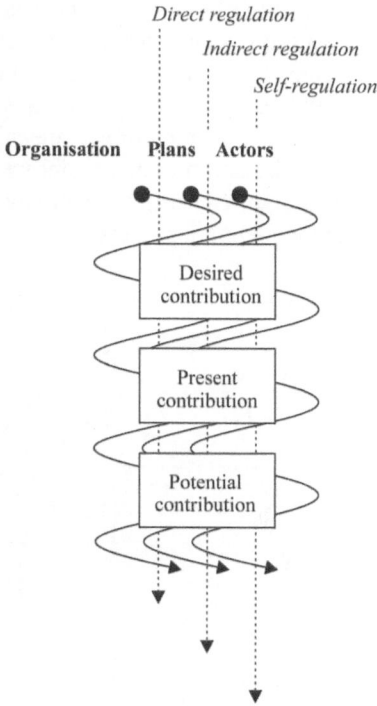

Figure 8.4 **The full representation of the policy model, including the focus on regulatory mechanisms within the process of planning and the attention to actors, parties and the plans they produce**

8.6 Actor-consulting and the various phases in the planning process

The outcome of the actor-consulting methodology provides a new vision for a regulatory framework, which reflects the key concerns of the main actors and elaborates them in a consensual way. This should result in the development of meaningful relations and collective actions. Scheff (1967), as already explained, takes the view that this is not necessarily a one-time event, which settles the case once and for all. Consensual understandings and arrangements are normally capable of lasting for only a short period of time, he argues. Obviously, actors collectively develop an understanding of the social structure that will frame their actions, but this remains a product of individual agreements and is always subject to renegotiation (Scott 1995: 109). Strauss et al. (1963) talk about a 'negotiated order'. The interaction can be structured through actor consulting in which meanings are produced, reproduced and transformed in support of meaningful relations and consensual actions. 'Once established in this way, collective actors may become agents for social action and are able to restructure themselves through collective learning processes' (Scott, 1995: 134).

The sociological argumentation teaches us about the need to be aware of the fact that once a collective agreement has been reached, it is susceptible to continuous change, which might end the consensus among the actors and their agreed course of action. According to Scott (1995: 104) there is a 'need for a constant monitoring of established meanings and the periodic creation of new meanings [to ensure] that the process of interpretation contains the ever present possibility of change' (Scott 1995: 104). In line with this argument, actor consulting might have to be repeated several times, at the various phases of the planning process.

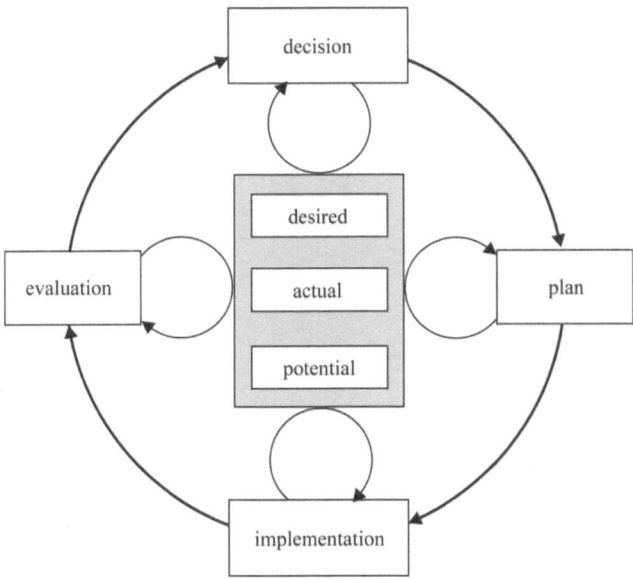

Figure 8.5   The planning circle and the actor-consulting model

If we use the planning circle (see Figure 8.5) as a representation of the various phases of the planning process, it would mean that actors should not only be consulted at the beginning of the decision-making process, but also when decisions are translated into a plan or programme. It might be necessary to consult the actors again when the planning process enters the implementation stage. It is possible, for example, that the move towards implementation of a plan brings with it a new set of actors, perhaps with a different attitude. This might be caused by various reasons, such as a misinterpretation of strategic language, a growing time span between the moment of decision and the implementation stage, or a dispute over technical details that brings unexpected barriers. Finally, there is an opportunity to consult the actors at the point when the effects of implementation are compared with the decisions made at the beginning of the planning process: the evaluation phase of planning. Here, a common understanding of the interpretation of events is needed as a basis for future co-operative developments.

In summary, the actor-consulting model is seen as a reflexive tool that can be used throughout the planning process. It is reflexive in a sense that it helps actors to confront the various aspects that come up within the planning process, with regard to their own actions, or the actions of others. However, as we have seen in Chapter 5, the actor-consulting model can also be used to be reflexive about the *information* that is used to present a critical perspective of the planning process (see Figure 8.6). Chapter 5 explained how the role of indicators might not only be used to reflect upon the planning process but also how we, as actors, should reflect upon the way we extract information from the planning process.

In conclusion, we should seek to understand the role of actors throughout the planning process. With this perspective we are able to take a more critical and reflexive approach to all the actions that we, together with others, make within any process of planning.

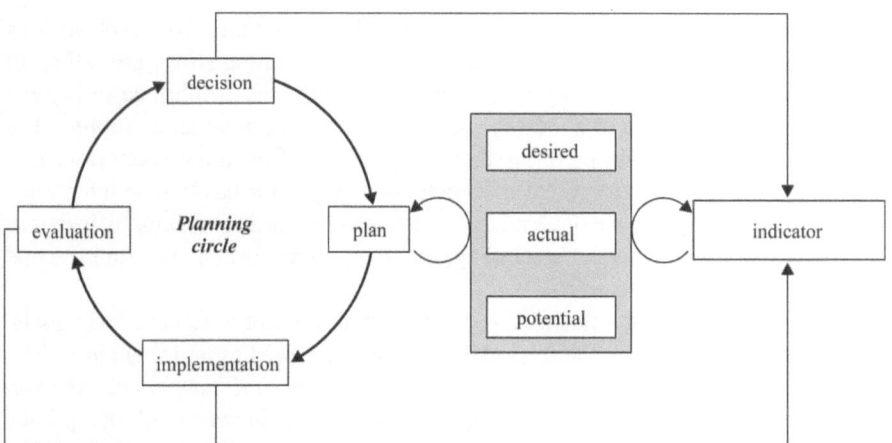

Figure 8.6   The planning circle, the actor-consulting model and the informative role indicators can play

8.7   Concluding remarks

We began Chapter 6 with a quotation by French that illustrates one of the most recognisable frustrations of planning. Despite all of our good intentions and thorough preparations, things turn against us as soon as an unexpected event occurs... We argue that full control is an illusion for most planning issues, that we have to be open minded not to focus on facts alone, but to also appreciate the opinions of the various actors involved. When planning issues become complex the approach in planning has to change accordingly. The approach will not necessarily become complex as well, but will nevertheless be different, involving a change in focus from an object-oriented towards an inter-subjective or institutional perspective. When relating this to the ongoing theoretical debate on planning it would mean a shift from a technical rational approach towards a communicative rational approach.

We started Chapter 7 quoting Lyotard saying that our social reality is neither determined nor stable, making it hard to extract a clear and common understanding of this reality. As such it becomes important to reframe our common terms of reference before we allow ourselves to take action. If we ignore the different perceptions of the various actors within the planning arena, certain aspects of the planning process might remain vague, fluid or fuzzy. If fuzziness is around – as it is likely to occur in complicated or complex situations – any attempt to reshape our environment through planning actions should be preceded by the reframing of the perceptions of the actors involved.

At the start of Chapter 8, Francis Bacon illustrates that 400 years ago, as nowadays, societal knowledge was taken for granted. In doing so we are failing to accept individuality, interactivity and subjectivity. Instead, we continue to believe that actors should behave in a way in which they are programmed, or supposed to act. It should be no surprise therefore that every so often frustration is our reward.

However, if we begin to accept that each and every actor has unique intentions, motives and perceptions, we have made a giant step forward. This book is trying to find a way to embrace this understanding, and this chapter brings us a tool to cope with this actor-related fuzziness.

We have argued that fuzziness is found whenever notions, concepts, goals, visions and desires are being addressed. Two aspects are seen as important here. The first concerns *understanding the underlying mechanisms that hamper the common understanding* of notions, concepts, goals and visions between all the parties involved, in order to improve the institutional setting and policy formulation. The second aspect refers to the *necessity of debate between the actors* about the meaning and definition of the notions, concepts, goals and visions that underpin the policy arena.

Considering the first aspect, the subjective nature of sustainability is used throughout this book as an example of fuzziness in planning. The subjective nature of sustainability stands in the way of a common understanding of how sustainability can be addressed. The proposed approach – actor-consulting – will bring insight not only to *what will be done* by the various actors but – more importantly – *why these actors are behaving in this way.* We will gain an understanding of the planning process by focussing on the *motives* of the actors involved. On the basis of such an analysis the objective of how to contribute to sustainability through planning could be rephrased into a more realistic set of objectives, which would be meaningful to the parties involved.

Considering the second aspect, the aim of the 'actor-consulting' model is to address the subjective nature of planning issues, to create a common understanding among actors, and to unravel underlying mechanisms that determine the actions of actors. This information equips planning authorities with a better anticipation in an uncertain policy arena, where many actors influence the decision-making process, and the outcomes are often unsure. The model aids planning authorities to formulate well-considered, 'realistic' policy by reducing this uncertainty. The application of the actor-consulting model is in fact a procedure to fine-tune the dynamics of the relationships between the various decision-makers and the other parties involved, and between the decision-makers and the executors of the decisions. Resulting policies are likely to be more realistic and more acceptable.   Actor consulting is therefore seen as a means to tackle uncertainties in planning that are not always recognised as such. In that sense actor consulting is a means to battle against prejudice in planning practice.

Actor consulting does not however guarantee that the actors will discuss the issues among themselves within a network. This feature lies in the realm of communicative planning, where the problems are more complex by nature, and where different planning tools are needed. The mode of governance served by actor consulting therefore tends to be one where an element of the top-down mechanism (see Chapter 3) is retained to some extent.

In Chapter 7 we have seen that fuzziness is likely to occur where 'complicated' situations begin to merge with 'complex' situations. In other words fuzziness can be found in those situations where efforts are needed to clarify interdependent relationships (complicated), or where uncertainty can no longer be avoided due to remote or unidentifiable cause and effect relations (complex). Obviously, in these situations both certainty and uncertainty are present, and neither a full technical-rational approach (certainty prevails) nor a communicative rational approach (uncertainty prevails) will suffice to address the situation. In Chapter 6 it is explained that when the content in planning is questioned a scenario approach can be used, while this chapter shows that in the case of actor-related fuzziness an actor-consulting approach might be an appropriate tool to grasp a better understanding of the issue at hand.

This brings us again to the spectrum of planning theory models, which we introduced earlier in Chapter 6. Our brief analysis results in four categories of decision-making in planning, the technical approach, the scenario approach, the actor-consulting approach and the communicative approach (see Figure 8.7). The *choice* of approach depends first of all on the complexity of a planning issue. Secondly, the *perception* of the planning issue's complexity is just as important. For example, one might be wrongly tempted to treat an issue as simple, because it appears to offer the prospect of full control. On the other hand, even simple cases might mistakenly be put under a regime of shared responsibility, relaxing the control at hand, when in terms of effectiveness and efficiency this would not be favourable. Therefore, a crucial third element that relates the planning issue with the categories of decision-making is the *context* of the issue.

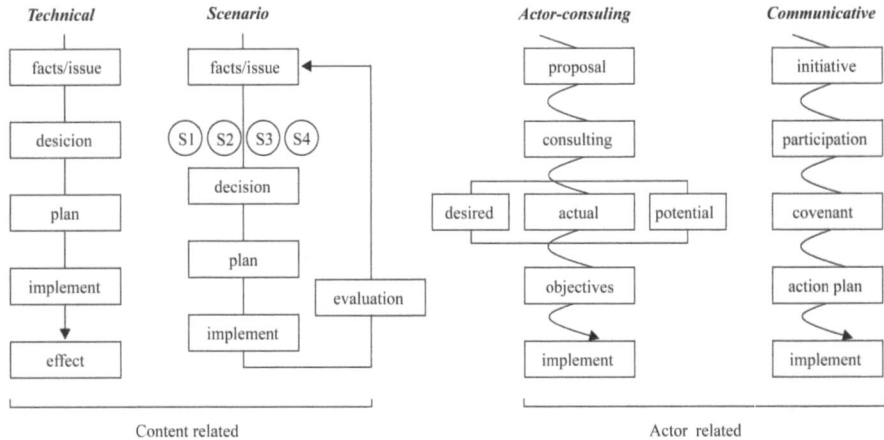

Figure 8.7   Planning processes for each of the four categories in planning

Choice, perception and context reflect strongly the subjective nature of planning, and from this perspective actor consulting could be seen simply as a practical tool to be used in every-day planning. However, it is seen here also from a theoretical perspective that deserves its own place in planning theory, filling in what we have identified as a missing link in planning practice and planning thought. Time will show what the arguments made in this chapter are worth. While we must await the evolution of these discussions, it is already clear that planners must continue to exercise critical thinking throughout the planning process in terms of what we want and why, in terms of what we do in practice, in terms of how we feel others should act, and so on. We have to be critical and reflexive to all the actions we make within any process of planning together with other parties involved. The actor consulting methodology offers a means to achieve these aims.

We sincerely hope the actor-consulting model can help us to gain a better understanding of how we can prevent 'bad things happening on beautiful days', and how we can avoid carelessness and overconfidence, with 'our gaze focused on the plan only', which is so often 'the moment when things start happening just outside [our] range of vision' (French 2002: 1).

References

Ackoff, R.L. (1979) 'The future of operational research is past', *Journal of the Operations Research Society of America*, Vol. 30, pp. 93–104.

Blumer, H. (1969) *Symbolic Interactionism*, Prentice-Hall, Englewood Cliffs (US).

Boje, D.M. (1995) 'Stories of the storytelling organization: A postmodern analysis of Disney as "Tamara-Land"', *Academy of Management Journal*, Vol. 38(4), pp. 997–1035.

CEA (1998) *Koersdocument sturing bodemsaneringsbeleid* [Guiding document soil sanitation policy], Rotterdam.

Cilliers, P. (1998) *Complexity and postmodernism: Understanding complex systems*, Routledge, London.

European Commission (2001) 'Directive 2001/42/EC on the assessment of the effects of certain plans and programmes on the environment', *Official Journal of the EC*, 21 July 2001, Brussels.

Giddens, A. (1984) *The Constitution of Society*, Polity Press, Cambridge, UK.

Hanf, K., I. Koppen (1993) 'Alternative decision-making techniques for conlict resolution: Environmental mediation in the Netherlands', *Policy and Governance in complex networks working paper no. 8*, Erasmus University, Rotterdam.

Harré, H.R. (1975) 'Images of the world and societal icons', in: K.D. Knorr, H. Strasser and H.G. Zilian (eds) *Determinants and controls of scientific development*, D. Reidel Publishing Company, Dordrecht (NL), pp. 257–283.

Kast, F.E. and Rosenzweig J.E. (1974) *Organization and Management: A systems approach*, McGraw-Hill Kogakusha, Tokyo.

Lane, D. (2000) 'Should system dynamics be described as a "hard" or "deterministic" systems approach?', *Systems Research and Behavioural Science*, Vol. 17, pp. 3–22.

Luhmann, N. (1982) *The Differentiation of Society*, Columbia University Press, New York.

MacIntyre, A. (1980) 'Epistemological crises, dramatic narrative, and philosophy of science', in: G. Gudding (ed.) *Paradigms and revolutions: Appraisals and applications of Thomas Kunh's philosophy of science*, University of Notre Dame Press, Notre Dame (US).

Rittel, H.W.J. and Webber M.M. (1973) 'Dilemmas in a general theory of planning', *Policy Science*, Vol. 4, pp. 155–169.

Rosenhead, J. (1996) 'What's the problem? An introduction to Problem Structuring Methods', *Interfaces*, 26(6), pp. 117–131.

Rosenhead, J. (2003) 'Problem Structuring Methods as an aid to multiple stakeholders evaluation', Paper presented at the Fifth International Workshop on 'Evaluation in Planning, Venice, February 14–15.

Rosenhead, J. (2005) 'Problem Structuring Methods as an Aid to Multiple-Stakeholder Evaluation', in: D. Miller and D. Patassini, *Beyond Benefit Cost Analysis: Accounting for Non-Market Values in Planning Evaluation*, Ashgate, Aldershot (UK), pp. 163–171.

Rosenhead, J. and J. Mingers (2004) *Rational Analysis for a Problematic World Revisited: problem structuring methods for complexity, uncertainty and conflict*, Wiley, Chichester (UK).

Scheff, Th.J. (1967) 'Towards a Sociological Model of Consensus', *American Sociological Review*, Vol. 32, pp. 32–46.

Schön, D.A. (1987) *Educating the reflective practitioner: Toward a new design for teaching and learning in the professions*, Jossey-Bass Publishers, San Fransisco.

Scott, J. (1995) *Sociological Theory; Contemporary Debates*, Edgar Elgar Publishing Limited, Cheltenham, UK/Lyme, US.

Snellen (1987) *Boeiend en geboeid: ambivalenties en ambities in de bestuurskunde*, Samsom H.D. Tjeenk Willink, Alphen aan den Rijn.

Stein, H.F. (1994) *Listening Deeply: An Approach to Understanding and Consulting in Organizational Culture*, Westview Press, Boulder (US).

Strauss, A., Schatzman, L., Erhligh, D., Bucher, R. and M. Sabshin (1963) 'The Hospital and its Negotiated Order', in: Freidson (ed.) *The Hospital in Modern Society*, Free Press, New York.

Visscher, K. (2001) 'Design methodology in management consulting', PhD thesis, Twente University, Enschede (NL).

Watson, S. and D. Buede (1987) *Decision Synthesis: The principles and practice of decision analysis*, Cambridge University Press, Cambridge (UK).

Ziegenfuss, J.T. (2002) *Organization & Management Problem Solving: A Systems and Consulting Approach*, Sage Publications Inc. Thousand Oaks (US).

Part C
Case Studies

Chapter 9

Sustainability through Information in the County of Viborg

Geoff Porter, Finn Larsen, Lone Kørnøv and Per Christensen[1]

9.1 Introduction

We now examine a case study based in rural Denmark, where countryside planning is founded on a system of spatial zoning. Planning applications are assessed and controlled in accordance with a pattern of functional uses that are set out on the County map. The zoning system has been progressively adapted over many years to accommodate increasingly complex and conflicting needs, in particularly for the protection of the environment. As in the UK case study (Chapter 11), a variety of policy appraisal and rationalisation tools are used in an attempt to minimise the conflicts and integrate new policy areas into local and regional plans.

Groundwater protection is of particular importance in the County of Viborg, where agricultural pollution easily penetrates the permeable sandy soils. One of the main planning issues resulting from this is the control of farming activities to prevent the pollution of drinking water aquifers, particularly with a view to facilitating their long-term protection. In granting planning permission for increased farming production, difficult choices have to be made between the socio-economic benefits of increased agricultural production, and their consequent environmental effects. This is not a problem that can be easily addressed by reference to the zoning ordinances. Consequently the need for more sophisticated planning tools has long been a matter of discussion between the planning officers and politicians within the County. The environmental planners at Viborg, with strong political support for the principles of 'sustainable development', had therefore taken the initiative to develop a pilot *internet-based* geographic information system (GIS), as a means of stimulating public and political debate. It was envisaged that this system would allow citizens – not least the farmers themselves – to model the possible effects

1 The research work for this case study was carried out by the Department of Environment and Technology at the County of Viborg (Finn Larsen), Denmark, with assistance from the Department of Planning and Development at the University of Aalborg, Denmark (Lone Kørnøv and Per Christensen), under the SUSPLAN project (SUSPLAN 2001a, 2001b), with the Sustainable Cities Research Institute, of Northumbria University, Newcastle upon Tyne as the lead partner (Geoff Porter). The project was part funded by the European Commission Interreg IIC North Sea Region Programme (Interreg 2001).

of increased agricultural production, using nitrate as an indicator of the extent of groundwater pollution. This pilot GIS system provides the background to our case study featuring actor-consultation.

The planners were interested to use actor-consultation to examine how to exploit the use of the GIS system, and to identify which actors, stakeholders and citizens should be encouraged to use it. On a more general note, the County also wished to explore how citizen participation in planning might be improved, and wanted to gather evidence to support the case for any further development of internet-based GIS indicators as a means to encourage participation.

In order to achieve these aims, it was decided to base the actor-consultation on an Environmental Impact Assessment (EIA) and the participative activities involved in that scenario. In this case, the EIA was linked to a particularly difficult planning application for the extension of an agricultural facility, which appeared to illustrate some of the problems and frustrations inherent in the existing planning system. As we will see, the conclusions of the case study support the need to develop tools to encourage self-regulation – aside from other means of regulation – as a means to prevent long-term groundwater pollution.

9.2   GIS as a means to develop public-political debate

As outlined above, it was apparent that activities of the farming community in the County of Viborg must be regulated in some way to reduce the potentially negative effects on the environment of continued increases in agricultural production. Our model in Chapter 8 illustrates that regulatory instruments can be categorised in three ways, and that a combination of direct regulation, indirect regulation, and self-regulation can be used to improve the outlook for 'sustainable development'.

After years of top-down and direct regulation, it had seemed apparent for some time to the environmental planners in Viborg County that the challenge in this situation was to facilitate an effective means of self-regulation. For a start, the citizens of the County needed convincing that there really was a problem, and further to this, they needed evidence of how their actions and behaviour might be affecting the situation. Armed with this information, citizens might be capable of contributing directly to 'sustainable development' in their own way, and on their own terms. The idea of using GIS was conceived in this context, based on the premise that a GIS presentation gives a more accessible and understandable overview than a presentation of statistics can normally ever hope to provide. In order to accommodate access by all citizens, the pilot GIS was established on the internet, via the website of Viborg County.

One of the Viborg County GIS presentations is shown in Figure 9.1. This illustrates the conflicts between the intensive agricultural production in the County and the quality of the drinking water. In the colour-coded presentations (presented in black-and-white in Figure 9.1), the GIS users can easily see where the quality of drinking water exceeds the standards set by the World Health Organisation. The level of Nitrate in water was chosen as the indicator,[2] supported by the fact that Nitrate is identified as an important environmental indicator both internationally and nationally (OECD 2000, UN CSD 2000). The nitrate concentration in surface waters and ground water has the advantage that it not only describes a condition that can be related to environmental and health standards, but also gives an indication of the intensity of agriculture. There is a close correlation between agricultural practice and leaching of nutrients to ground water and watercourses.

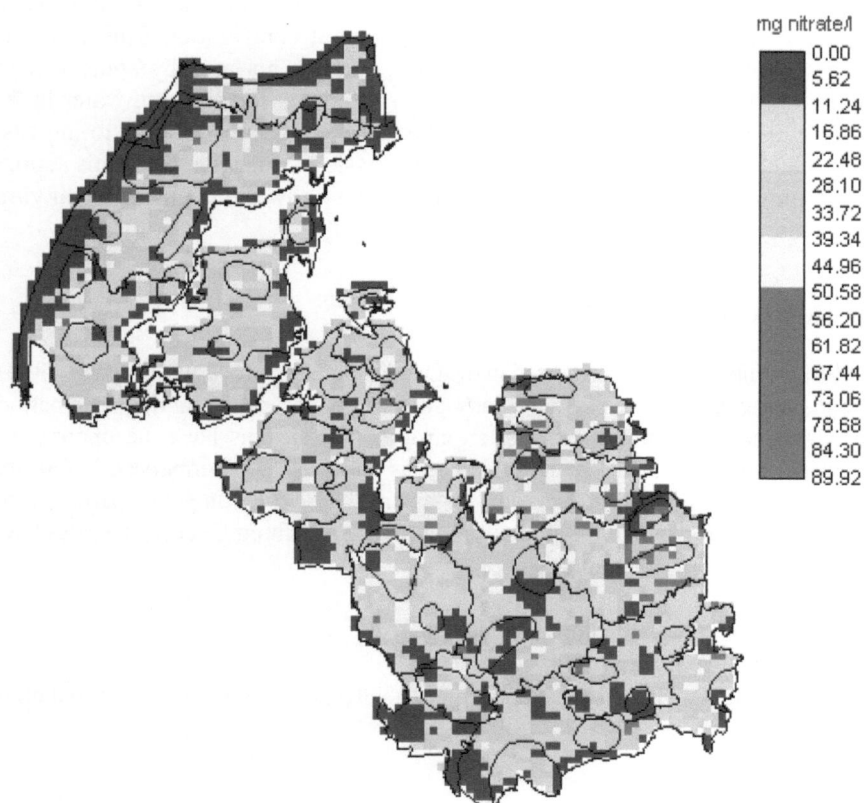

mg nitrate/l

	0.00
	5.62
	11.24
	16.86
	22.48
	28.10
	33.72
	39.34
	44.96
	50.58
	56.20
	61.82
	67.44
	73.06
	78.68
	84.30
	89.92

Figure 9.1   GIS presentation illustrating the concentration of nitrates in recently-formed groundwater, juxtaposed against areas designated as having special drinking water interests (indicated by thin black outlines), across a map of Viborg County (Viborg County 2001)

2   A calibrated mathematical model (based on empirical measurements of the nitrate indicator) is used to present a colour-coded picture of the nitrate levels in water, using parameters selected by the user.

It was anticipated that this type of presentation should attract the attention of members of the public especially in areas with special status for drinking water quality. The aim was to allow different scenarios to be modelled by citizens, in accordance with their own choice of location and other selected parameters, such as the number of animals per hectare. Effectively, the user can see in the first instance a pictorial view of what is effectively their *present contribution* to sustainability, and can go on to explore the outcome of their *desired and potential contributions.* Here we are observing, *before we get into the actor-consulting methodology*, that the GIS gives citizens an insight into their own self regulatory contributions. For example, a potential contribution from a farmer might be an initiative to reduce the nitrate load in his area by some means, leading to the achievement of the minimum standards for drinking water in the vicinity of his farm. By way of a second example, a farmer wishing to increase his production can use the GIS to model the possible consequences of his actions on groundwater, and therefore gain an insight into how the planners might view his planning application.

9.3   Aims of the case study

Public participation was introduced in the Danish planning process in the early 1970s. The Planning Act includes requirements such as openness and public participation. The first phase of public participation ensures that citizens have the opportunity to debate the scope of the plan, submit ideas and propose alternatives, before the County Council decides upon a policy statement. These rules on public participation are the minimum national requirement. In the case of Viborg County, the rules have been extended with, for example:

- dissemination of material for debate;
- public meetings;
- meetings with actors with a special interest in a case (e.g. a planning applicant);
- study trips;
- focus groups;
- demonstration days;
- contact between politicians and citizens.

Planning is based on a participatory tradition of bringing various stakeholders or interest groups together in order to advise the authorities. Viborg is now preparing its sixth generation of Regional plans, so experience with public participation is fairly comprehensive. However, there is still potential for improvement.

In 1998–1999, an analysis of Regional and Municipal planning in Denmark was undertaken and several problems relating to the involvement of the public were identified (Ministry of the Environment and Energy 1999). Firstly, planning had to be presented as more relevant to the public in order to attract the participation of citizens. This pointed to the need to encourage participation based on more local or specific matters. Another problem raised was that the authorities had a tendency to maintain contact primarily with the well-established groups. Even though many citizens can be represented through different organisations, individuals were regarded as having a weak position. The actor-consulting case study therefore sought to develop a picture of how citizens were motivated to take part in the planning processes, with a view to developing an insight into how citizen participation might be more effectively accommodated.

9.4   Using actor-consulting to examine participation and communication

The need for the Environmental Impact Assessment of planning applications for increased production capacity on farms has increased, as Danish pig production has increased from around 18 to 23 million pigs per year during the 1990s. EIA is seen as a means of assisting planning decisions, by ensuring that the environmental effects of increased production are controlled. In many locations (especially designated drinking water zones) such an increased production may violate the targets for ground water protection. This leads to conflicts, although hitherto many of the Danish Counties have approved applications for increased production within these areas.

The actor-consulting study was based on the EIA of a proposal for the extension of production at a pig farm. This case was chosen for several reasons. Firstly, it was related to the indicator for water quality that had been chosen for the GIS pilot study. Secondly, farm production was high on the political agenda in the County, and thirdly the planning application had continued for an unprecedented four-year period, and in doing so had raised many concerns. In many respects the chosen case study therefore represents an extreme case, allowing us to infer how the question of sustainable development was handled under conditions of conflict between the involved parties.

The research relied to a large extent upon a study of the EIA documentation, supported by interviews with the actors involved in the EIA case (see Figure 9.2). The interviews were planned and conducted by researchers from the University of Aalborg. The information derived from the actor discussions was used to qualify the documentary study and to obtain the opinion of the actors with particular reference to: the planning process in general; their perception of the way in which the EIA had been conducted; the extent and nature of each actor's participation; the motivation of each of the actors; the extent of their dialogue with the County; and how participation might be improved.

Table 9.1   Actors consulted in the Viborg County case study

National, Regional and Local Government Departments
   National Forest and Nature Agency (Ministry of the Environment and Energy)
   Spatial Planning Department (Ministry of the Environment and Energy)
   The Commission for Agriculture
   Viborg County (the planning authority)
   Karup Municipality

Business interests
   The farmer
   Agricultural organisations

Community / Environment interests
   Individuals involved in the EIA public hearings
   The Outdoor Council

9.5   Outcome of the actor consultations

The study enabled the planning authority to develop an independent, 'eye-witness' review of the planning process that had previously taken place under their jurisdiction. The outline of the story revealed by the actors is set out below, illustrating the complexity of the technical arguments, and the complex interplay between the interests of the actors, citizens and politicians that took place as the EIA progressed. This story shows the *present contributions* of the actors, although it is of interest that their desired contributions are often hidden from the view of the planning authority.

The farmer in question lives close to Viborg, in a sandy area with exposed groundwater aquifers, and where no layer of clay is in place to offer protection. The groundwater in this area is abstracted for use in the nearby towns. In 1996 the farmer applied for an increase in his production from 4500 to 7100 pigs per year. The animals are accommodated indoors, and steps had already been taken to process a proportion of the slurry by means of bio-gas production. Consequently mitigating measures were already in place with respect to the sustainable reduction of nitrates.

In the first instance, the County considered that the planning application was unlikely to be problematic. However EIA was still a new instrument at that time and an element of confusion regarding the context of its use was present in most of the government hierarchy. Consequently, it was nearly a year after the initial planning application that it became clear that the County would have to initiate the EIA procedures. This commenced with a public consultation, where a discussion of ideas and a general debate took place. During that summer the County stated that they found the area vulnerable to farming activities, and published a leaflet inviting the public to take part in discussions and meetings, and to object to the proposals if they found that appropriate. The Danish Nature Conservancy Association objected to the project on several grounds. Their local committee in Karup Municipality underlined that it was problematic not only in relation to groundwater protection but also that the release of ammonia to the air could harm nearby areas protected under the law of Nature Protection. Besides this, a group of neighbours objected to the increase in production, also using the argument of the protection of groundwater, but furthermore questioning if smell and flies would increase as well.

Based on these objections, the County administration decided that they could not approve the increase in livestock production. However, a number of politicians did not accept the recommendation to reject the application. In order to make the relevant decisions the County requested new information on annual farm production, and the application process started again. This time the County administration decided that they would use a quantitative model, developed by the Danish Nature Protection Agency, to assess the viability of the planning application. During the summer of 1998 several calculations were made, which concluded that the resulting nitrate concentration at that time was slightly above 100 mg/l, which is considerably higher than the maximum limits set at the national level for drinking water. The calculations indicated that the level of pollution would decrease as an effect of the general regulations to be implemented in the years to come, but the proposed increase in production would again raise the level to around 100 mg/l. Armed with this information, the County administration again put forward that the increase in production could not be allowed, but stated that if major changes in the crops grown on the farm took place (more grass and pasture) then this might serve to absorb the load of nitrogen sufficiently. Consequently, the politicians asked the administration to continue the EIA process.

In the autumn of 1998 the second public hearing took place. It was stated that the increase in production could only be allowed if it could be firmly illustrated that the leaching of nitrate would not exceed the guideline limit. This time comments and objections from two groups of neighbours, as well as the Nature Conservancy Association and Friends of the Earth were submitted to the County. The objections were on the grounds of lack of groundwater protection, the need for protection of natural areas from ammonia, smell and air pollution, and also upon the grounds that the proposals would lead towards a less varied landscape. Basically, we could say these are an expression of *desired actions* presented by the advocacy groups.

As part of the public consultation, the farmer himself queried the legal foundations for imposing such strict limits on his production. At that point in time, the Danish national regulations on groundwater protection were very new, and the interpretation of the regulations in the context of 'nitrate vulnerable' areas was questioned. The accuracy of the mathematical model used to evaluate the nitrate load was also questioned.

Shortly after the public hearing the County was approached by several agencies under the Ministry of the Environment, probably influenced by the Farmers Union, who raised the question of how to re-formulate the *targets* for nitrate concentration in water. The Ministry and its agencies sat down to discuss the situation, and in the spring of 1999 ruled that the targets used so far were legal, although they could not be applied in the case of neighbouring areas where the farmer had arranged to dispose of the slurry. They also ruled that the model used was the only one that at that time was appropriate for this purpose.

In the spring of 1999, three years after the case began, the County initiated discussions with the farmer and his consultant on how to make these calculations. The farmer said that the limits put forward in the EIA would prohibit any normal agricultural practices on his fields. The discussion centred on how to calculate the exact limits. The farmer and his consultant argued that not only the concentration of nitrate in the water leaving the root-zone of the field should be calculated, but that the capacity of the soil beneath to absorb the nitrate should also be taken into consideration. Furthermore they underlined that the lower nitrate load coming from forest and natural areas within the designated drinking water zone in question should be taken into consideration as well, as the trees would have the effect of attenuating the seepage of nitrate into the groundwater. These issues can be considered as the *desired actions* as proposed by the farmer.

The County administration initially took the view that nitrate reduction capacity should not be taken into consideration in this way, as this capacity would be used up in the years to come. During the spring of 2000 the County administration changed its view on nitrate reduction capacity, and now stated that it could be used in the calculations. Consequently, only minor changes in agricultural practices (e.g. a larger percentage of grassland) would be needed to accommodate the quantities of nitrate put forward in the EIA proposal. For the farmer this meant that he should change his production so that 30% of his fields should be grown with spring barley, and then later replanted with grass in order to capture some of the nitrate during the autumn. In mathematical terms this was feasible, but nevertheless the County administration decided not to approve the increase in production, arguing that as the requirements for crop planting could not be imposed on the neighbouring farmer's fields to which he delivered slurry, the impact on the groundwater could not be controlled.

9.6   Discussion

Regulatory instruments

The actor-consultation has therefore revealed a rather technically oriented decision-making process that features many uncertainties. The situation appears at first to be very simple, but soon gets bogged down in complex technical arguments. However, much of this argument might with hindsight have been simplified if all of the parties involved had had a common platform of understanding, and in the case of most citizens, a better understanding of the specific linkage between increased agricultural production and drinking water quality. This led the planning authority to believe that the GIS system would have been effective in supporting this process, providing that all of the actors, stakeholders and citizens had had access to the same information. Public access to the GIS was therefore considered to be of importance.

Self-regulation might at first glance appear to be ineffective in a situation such as this, which is typified by technical argumentation and conflicting interests. The GIS does not constitute a system of self-regulation in itself, but may help to facilitate a public and political debate, which might – or might not – in turn lead to individual self-regulatory contributions by the farming community to complement the existing direct regulatory regime. Of course, other regulatory options such as indirect regulation are possible, and further consideration could be given to this type of approach. Nevertheless, bearing in mind that the planning system currently relies very much on direct regulation and does not give clear and acceptable outcomes, any move towards a complementary self-regulatory framework would appear to be desirable. The GIS therefore represents a tool to facilitate this move.

Transparency and simplicity?

The case study has exposed many difficulties inherent in pursuing EIA that relies upon a technical argument, based on mathematical models and pre-defined scientific indicators, where the questioning of the interpretation of results, and arguments about the perceived 'fairness of the rules' can appear to go on forever. The EIA process was limited to issues of environmental pollution, and did not seek to explore the social / economic dimensions of the planning application. The terms of EIA do not readily accommodate a study of the effect of refusing planning permission upon local jobs and the local economy. It may be that the Strategic Environmental Assessment Directive (European Commission 2001) prompts national legislation to call for the 'sustainability appraisal' of Regional Plans (Kørnøv 2001), and therefore a broader range of considerations might be possible in evaluating future planning applications for farming expansion. Consequently, it will be of interest to Viborg County to explore indicators linked to objectives based on social and economic considerations, and their interaction with a range of environmental parameters. Work of this nature is already underway.

As anticipated, the actor-consultations supported the notion that GIS presentation of indicators will help to support the implementation phase of the planning cycle, by raising awareness of a range of sustainability issues – in this case – water quality. As a by-product of this, the process of EIA and its accompanying political debate can be rendered more transparent by using GIS as a stimulus for debate throughout society. A potential danger here, however, is that the interested parties do not agree on the selected GIS indicators, and this underlines the importance of securing participation at the onset of the process, and the importance of ensuring that the indicators represent information relevant to the stakeholders involved. (Chapter 5 suggests that indicators might be an *output* of an actor-consulting process.)

In a parallel study of planning processes in a neighbouring County, where environmental assessment was integrated into the development of the regional plan, a County Council politician expressed: 'It is not clear what weighing up (of alternatives and interests) took place during the planning process. So you may worry about how the interests were handled. It would have been good to see the basis on which the decisions took place.' (Kørnøv 2000). Another politician expressed the view that: 'The difference in the level of knowledge between officials and politicians is considerable. Over many years, officials have developed a way of writing, which makes it difficult to recognise the actual problems. We need to find a more precise way of describing environmental assessment and a common level (of communication). I think this will improve future discussions considerably.' (Kørnøv 2000). Monitoring of public reaction to indicators, and their graphical representation, may in future help to address these questions. The use of GIS may therefore help to qualify the political debate about the preparation of plan policies.

We could summarise by saying that the Danish case study has attempted to move the focus of public participation in planning processes from a 'suppressed dialogue' to a 'qualified and meaningful dialogue'. In other words, it is not only a question of allowing a formal opportunity to participate. It is indeed a question of further transparency and simplicity for both the politicians and the public.

Public participation in EIA and regional planning

As can be seen from the story in Section 9.5 above, the actor consultation revealed much information about opinions and motives. The concern of the farm's neighbours tended to be focused on superficial issues, such as smell, flies and an aversion to changes in their neighbourhoods. Of course, the planning authority takes these concerns seriously, but they do not in this case appear to deliver any central arguments to the question of how the expansion of the farm might affect water quality.

The interviews revealed that public participation in this type of local issue can be difficult. The public who responded tended to be newcomers to the local community and therefore they felt freer to respond in the public hearings. Conversely, many established local people decided not to let their voice be heard in the process, even though they did not support the extension of the farm production. Their *desired contribution* was therefore hidden from the view of the planning authority. This points to the challenge of arranging public hearings in a way that overcomes this barrier in small communities. Another point raised was the importance of having a formal response from the authorities making clear firstly that the County has received an objection from each responding citizen, and secondly how the objections have been handled and the reason for the final decision. In summary, the actor-consulting revealed that the EIA / planning application process was failing to maintain a sufficient level of communication with objectors.

Most of the actors in the formal hearing were public authorities and larger Non-Government Organisations. These actors tended to deliver coherent arguments on how to regulate nitrate pollution, and were capable of addressing local issues in the context of the broad discourses at the societal level. Contributions from local people tended to lack this focus on the question of water pollution.

Unsurprisingly, the case study concluded that all of the involved parties considered that the EIA process had taken far too long. The reason for this may be that this was one of the first EIAs on farm expansion, and that the new regulatory measures were not always clear. During the process, Viborg County was actively involved in discussions with central government in order to clarify the rules, and how to handle them in a specific case. Some of the involved parties consequently felt that the County had problems with decision-making. As the case shows, though, the 'truth' about the nature of the problems and how to calculate the impact is always negotiable. And in this case the negotiations took place at a very central, national level involving the Ministry and its agencies as well as actors within the field of agriculture. In this sense the process of establishing what 'sustainability' is took place among central partners and was thus more of a top-down nature than bottom-up.

Some of the involved parties felt that the decision to reject the application for the enlargement of the farm was a foregone conclusion. However, many of the arguments that the County integrated into their decision favoured the farmer in the sense it became more and more obvious that it would be possible to comply with the environmental limits by introducing only small changes in crop management. Instead, the argument that neighbouring fields for the deposit of slurry could not be regulated became the crucial argument for rejecting the proposals for increased production. For many of those involved this seemed a far-fetched argument and therefore they felt that the result was determined in advance. In a sense though, this shows a misunderstanding of the intentions of the planning authority. This returns us to the need for tools that support transparency, which again supports the need for GIS, as a means to promote a better-informed public.

9.7 Concluding remarks

The actor-consulting study of participation in EIA of farm projects[3] concluded that the concerns of the public should be taken more into consideration and that this would help to reduce tensions and conflicts. The case illustrates the need for more accessible guidance on the nature and extent of water quality problems in the County, and it is hoped that the embryonic GIS indicator system might help to fulfil this need. The GIS does not attempt to deliver certainties, in terms of predicting or deciding if groundwater policy is 'sustainable', but will simply assist in building dialogue and public understanding. An ongoing dialogue with the key actors will continue, as a basis to monitor the uptake of the GIS facility and its effect in promoting an effective public-political debate.

The role of actor-consulting has been effective in clarifying the disadvantages and limitations of the formal EIA system. A better understanding of the role of actors and citizens in the planning process has been achieved, and this will assist

---

3   Note that the case identifies quite closely with 'complicated' cases, as discussed in Chapter 7.

in accommodating their needs more effectively in future generations of the Viborg Regional Plan.

References

European Commission (2001) 'Directive 2001/42/EC on the assessment of the effects of certain plans and programmes on the environment', *Official Journal of the EC*, 21.07.2001, Brussels.

Interreg (2001) *Interreg IIC North Sea Region: The Projects, Programme Secretariat*, Viborg (DN).

Kørnøv, L. (2000) 'Strategic Environmental Assessment in an imperfect world – a study of rationality in planning and decision-making processes', Ph.D. Thesis, Aalborg University, Aalborg (DN).

Kørnøv, L. (2001) 'Integration of strategic environmental assessment in spatial planning: the Emperor's New Clothing? SUSPLAN 2001 – The transformation to sustainable planning: decision-making, models and tools', Conference Proceedings. Sustainable Cities Research Institute, Newcastle upon Tyne (UK).

Ministry of the Environment and Energy (1999) Strategy and municipal planning, London.

OECD (2000) 'Towards Sustainable Development, indicators to measure progress', proceedings of the OECD Rome Conference, Paris. See: http://www.oecd.org/env.

Susplan (2001a) *SUSPLAN – Developing tools for sustainable planning, A Common Approach to sustainability in spatial planning*, August 2001, Newcastle Document Services, Newcastle upon Tyne (UK).

Susplan (2001b) *A Common Approach to developing sustainability in the Spatial planning processes of Denmark, Netherlands and the UK.* (Issue 7; September 2001), Sustainable Cities Research Institute, University of Northumbria, Newcastle upon Tyne (UK).

UN CSD (2000) 'Indicator sets by the UN Commission for Sustainable Development', Division for Sustainable Development, Department of Economic and Social Affairs, United Nations, New York, N.Y. 10017, U.S., made visible at:www.un.org/esa/sustdev/isd.htm.

Viborg County (2001) *Regn din egen udgave af nitrattabsmodellen*, Aalborg (DN), made available at: http://gis.vibamt.dk/nitrat/.

Chapter 10

Policy Life Cycles: Cultural Heritage in the Dutch Wadden Sea Region

Rob de Boer[1]

10.1 Introduction

A sustainable use of historical structures and features (ancient monuments, archaeological artefacts and valued landscapes) is one of the main goals of the Netherlands' cultural heritage policy (OCW et al. 1999). From the features and structures of the past we learn about how people used to live, and we gain knowledge about their cultures and habits. By preserving them, future generations will also be able to learn something from their past. Hence, the notion of 'sustainability' in our research of the use of the Dutch Wadden Sea Region has had a strong temporal and preservation dimension. This chapter is a reflection on how the actor consulting model has been used to guide this review of cultural heritage policy (see also De Boer and Jager 2001).

10.2 Cultural heritage in the Wadden Sea Region

The Dutch Wadden Sea Region (see Figure 10.1) is well known for its special values in the field of cultural heritage, both at the national and international levels. It is the Dutch part of the greater Wadden Sea, which falls under the jurisdiction of three countries: Denmark, Germany and the Netherlands. The research described in this chapter was carried out in the Dutch sub-project of the EU-funded LanceWad-project, which covers the greater area of the Wadden Sea.

The Wadden Sea stretches from Den Helder in the Netherlands to Esbjerg in Denmark, and also covers areas in two German *Bundesländern* (Lower Saxony and Schleswig-Holstein). For more than 2500 years people have lived and worked in this area. The struggle against the elements of the North Sea has over the centuries transformed the natural landscape into the man-made landscape we know today. The mounds and dykes, the reclaimed land and fishing villages, the old towns and lighthouses, the pattern of the wooded banks and ditches are the visible evidence of this transformation. Outside the dykes lie the silent remains of bygone activities,

---

1   Rob de Boer is researcher at the Faculty of Spatial Sciences, University of Groningen, the Netherlands.

including wrecked ships from the Dutch Golden Age or even Viking relics. Together
they tell the history of the Wadden Sea Region. The quality and unique character of
the historical structures and features of course make it worth preserving them for
future generations. The international project LanceWad was established to address
three main aims: to make an inventory and carry out an evaluation of cultural
heritage; to research the perception of the environment; and to promote sustainable
use of the cultural heritage.

It was for this last aim, that the Dutch Ministry of Agriculture, Nature and Fishing
wanted to research the desired, actual and potential contribution of the policy-making
actors concerned with cultural heritage in the Dutch Wadden Sea Region.

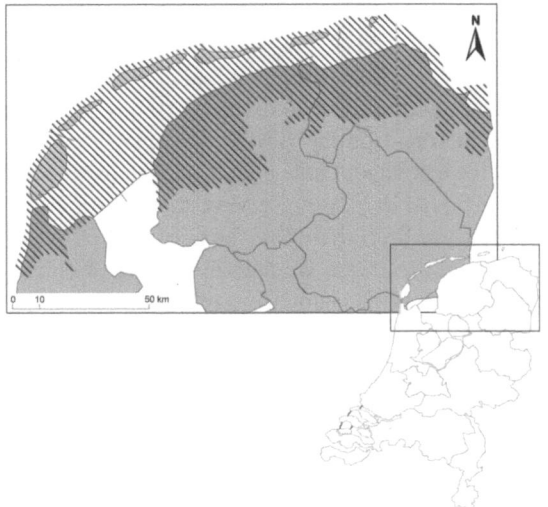

Figure 10.1   The Dutch Wadden Sea Region

10.3   The cultural heritage policy arena

The number of UN World Heritage protected areas is still rising and the European
Union has special programs on regional identity and cultural heritage. This trend
is also visible from a Dutch national perspective. In the first half of the 1990s the
national government incorporated aspects of cultural heritage into the Dutch National
Spatial Plan (VROM 1999). In its successor, the dimension of cultural heritage and
cultural identity received an even more prominent role and is seen as one of the
seven main qualities of space identified within Dutch spatial planning (VROM
2001). Obviously, attention to cultural and societal heritage is rising alongside the
growing recognition of the qualities of the Wadden Sea Region.

The Dutch central government also drew up a special memorandum on the
relationship between cultural heritage and spatial planning, the *Nota Belvedere*

(OCW et al. 1999). The central subject in this memorandum is the relation between preservation of cultural heritage and future spatial development. This relationship has traditionally been under pressure, resulting from the notion that preservation of cultural heritage limits the possibilities for spatial development and that future developments could affect the quality of cultural heritage. The Nota Belvedere breaks with this restricted school of thought, and takes the relation one step further. Here it is argued that the preservation of cultural heritage can support spatial development and vice versa (i.e. *preservation through development* and *development through preservation*). Preservation and development are no longer seen as two conflicting issues, but rather as two mutually strengthening sides of the same coin. So a good balance between preservation and development can provide added value to the quality of our environment.

10.4 Using the concept of the 'Policy Life Cycle'

An important consequence of this recent policy shift was that, as discovered during the interviews for this research, the possible effects of the new policy guidance (involving for example changes in areas of responsibility between the actors, or the introduction of more specific policy instruments) had not yet filtered through into practice. The implementation of the new policy guidelines appears to have taken place on a rather ad hoc basis, instead of providing structural change with respect to the advancement of cultural heritage. On the other hand it could be seen that the expectations were high and the belief in this new policy was strong. This was experienced through the numerous examples and success stories given throughout the interviews.

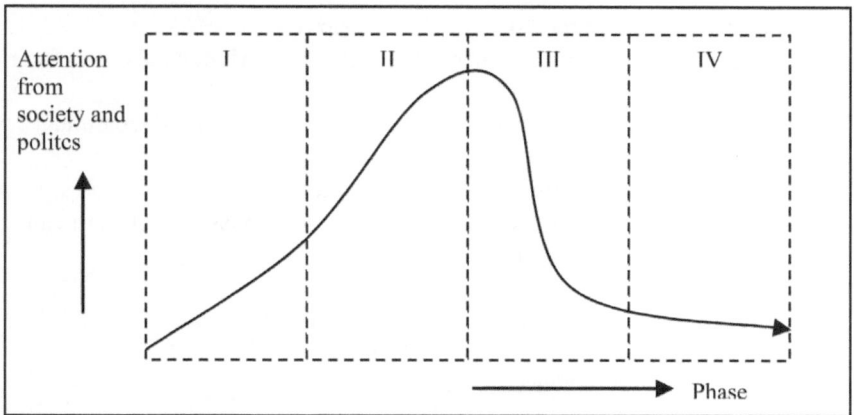

Figure 10.2   The policy life cycle as proposed by Winsemius (1986)

In summary, all of the actors without exception were willing to contribute to more sustainable practices regarding heritage and spatial planning, but were unsure how to take the situation forward. It was felt that instead of talking about their present contribution the interviewees talked about very basic issues with a wide scope or even in some cases about their 'desired' present contribution in future times. A tool was needed in this situation to analyse the present contributions to sustainability in a more structured way. With this in mind, a 'Policy Life Cycle' was introduced to model the present contributions against different stages of the cycle (see Figure 10.2). A policy cycle with four consecutive phases was used as follows: 1.) The recognition of the policy problem, 2.) The formulation of policy goals, 3.) The implementation of policy goals, and 4.) The management of the policy approach (see Winsemius 1986).

These four phases are explained in more detail below, and the results of the research are presented according to this classification.

10.5   The actors

In order to collect information about the desired, present and potential contributions, the researchers visited over thirty organisations. The actors had a broad composition, including staff members of the municipalities and provinces, mayors, members of the Provincial Executive of the three provinces in the Dutch Wadden Sea area. The actors are summarised in Table 10.1.

Table 10.1   Interviewees participating in the actor-consulting project

---

Employees from government departments
  The Staatsbosbeheer (Dutch Forestry Commission);
  The Rijksdienst voor Monumentenzorg (Department of Monuments and Historic Buildings);
  The Rijksdienst voor Oudheidkundig Bodemonderzoek (Department of Archaeological Soil Research);
  The Rijkswaterstaat (Department of Public Works);
  The Departement van Landbouw, Natuurbeheer en Visserij (Department of Agriculture, Nature and Fishing).
Delegates of the water boards
  People from interest and conservancy groups including:
  Waddenvereniging (Wadden Sea Association);
  It Fryske Gae (Friesian Landscape Council);
  Stichting Groninger Landschap (Groninger Landscape Council);
  Milieufederatie (Federation for Environmental Protection).
  Representatives from user groups such as:

Noordelijke Land- en Tuinbouw Organisatie (Northern Agriculture and Horticulture Organization);

The Midden en Klein Bedrijf (Association of Medium and Small Enterprises);

Agrarische Jongeren Kontakt Friesland (Association of Friesland Agricultural Youth)

Interviews with these people, as individuals and as representatives of their respective organisations, did not use a predefined definition of sustainability, or a predefined set of sustainability indicators. This would contradict the principles of the actor consulting model. Instead an open discussion took place about sustainability and cultural heritage with simple guidelines for the discussion topics. As a consequence of the open discussion the scope of the study was as wide as the interviewees made it, but the bulk of the discussion was about the available policy instruments and the implication of these instruments in relation to the sustainable use of cultural heritage. The notion of desired, present and potential contributions to sustainability served as a structural element in the interviews.

10.6 The desired contribution of actors to sustainability

The research indicated that there was a great deal of agreement between the actors involved in the decision-making arena of the Dutch Wadden Sea Region. The desired contribution to sustainable management of cultural heritage was consistent with the approach of the Nota Belvedere, and also with recent developments in Dutch policy on decentralisation, deregulation and market-orientation.

There was a strong individual desire among the actors to contribute to the sustainable management and use of cultural heritage. The actors wanted a decision-making process involving clear-cut task descriptions, commitment to action, social support, close consultation, and exchange of information between the parties involved. The desire was to come to a good balance between preservation and development, as was intended by the Nota Belvedere. For instance the rigid division between financial schemes supporting either preservation or development is now replaced with a financial scheme supporting projects with a strong integrative nature. The idea of using cultural heritage as an inspiration for new spatial developments was seen as a progressive approach to policy making. So with regard to the desired contribution, it would be easy to conclude that every actor was supportive of the sustainable use of cultural heritage in the Dutch Wadden Sea Region and that a bright future would be ahead of us. However, on some points there were major gaps between the desired and the present contribution, so that here the positive outlook took on a somewhat different complexion.

10.7   The present contribution of actors to sustainability

Recognition of the policy problem

In this first phase of the Policy Life Cycle, a certain theme will generally attain social importance or priority. The crucial question in this phase is whether the theme is in fact really a policy problem in need of attention and steering from the government, or whether self-regulation can be left to take its own course. In the case of cultural heritage, the theme has gained attention throughout society in the last ten years: transforming it into a real policy problem and the appearance of the Nota Belvedere could be seen as a result of this. Our research reveals that cultural heritage is seen as an important theme in holistic town and country planning.

The actors in the decision-making arena in the Wadden Sea can be characterised as broadly recognising the 'problem' of cultural heritage, and as very willing to contribute to its sustainable management and use. Even the actors with no direct interest in cultural heritage, such as those from the agricultural sector or economic enterprise were willing to recognize the need to contribute to the sustainable use of the cultural heritage, because it was seen as a major part of their own individual culture and history. Now, the key issue becomes whether this broad recognition of the policy problem might firstly be supported by the formulation of policy goals, and how these goals might subsequently be translated into concrete action.

Formulation of policy goals

The aim of this second phase in the Policy Life Cycle is to involve actors collectively in the formulation of policy goals. These policy goals can take the form of a spatial plan, a memorandum, or established guidelines. In participating in the formulation of goals the actor commits to providing some form of contribution that will assist in transforming policy goals into action. The provincial plans of two of the three provinces of the Dutch Wadden Sea Area now include such explicit goals in their plans. Also the municipalities of the five Wadden Sea Islands have developed a special memorandum on cultural heritage. Opposite to this albeit slow process of the formulation of goals, the Wadden Sea research found that the cultural heritage theme had not been embedded into organisational structures. Many organisations and especially the smaller ones had no staff member or employee with a clear responsibility to formulate their policy and to develop their role in the cultural heritage theme. Positive contributions were usually represented by the actions of a dedicated member of staff, but this was clearly not yet recognised as a structural task. For instance with the smaller organisations it was sometimes hard to find the right person to interview at all. This under-resourcing of the cultural heritage issue stands in contrast with the perception that cultural heritage is seen as an important priority. Nevertheless, it was concluded that although the progress was slow, the future appeared to look promising.

Implementation of policy goals

The third phase of our policy life cycle concerns the implementation of the goals. The dividing line between phases two and three lies in the translation of policy goals into concrete tasks. The formulation of goals therefore is an important precondition for their implementation. The Wadden Sea actors involved in the implementation argued consistently that the present way of formulating goals is too vague and too strategic. The policy goals were generally formulated in a phrase like "For questions of spatial design we should take into account the environment, nature, the water balance and last but not least our cultural history". However there were no guidelines as to how each of these aspects might be weighed relative to each other, or how to act in the case of conflicting interests. There were insufficient concrete guidelines on what to do and in particular on how to get started. Consequently, the implementation of cultural heritage was in practice typified by ad-hoc initiatives that made little reference to any structural approach. For instance, there were no examples of urban expansion paying attention to cultural values. Furthermore it seems that the question of *priority* for implementation was guided by economic good, instead of cultural values. The need for preservation of cultural heritage in combination with growth of restaurants or shops was frequently named as an example, as well as the need for interpretation facilities and information centres. So, based on the reactions of the actors, it can be concluded that there is a clear and widening gap between the desired contribution and the actual contribution in the implementation of policy goals. This situation can only be addressed by the development of a more purposeful set of policy goals for the sustainable development of the Wadden Sea Region, that relates to the expectations of a wide range of actors.

Management of solutions

The fourth phase of the policy life cycle is the *management* of solutions, consisting of inspection, maintenance, and aftercare of policies, agreements and projects. The management phase has in the last ten or twenty years received insufficient attention according to the actors. The biggest problem of this phase appears to be that it currently enjoys very little in the way of public and political support. Political attention is focused on new issues, leaving the old issues with less attention, and this in turn leads to a lack of economic support for the management of cultural heritage issues. The actors remarked that in the management phase, the actual contribution is nowhere near the desired contribution. An example of this is that a lot of recently recovered archaeological artefacts have received no attention since their discovery. Another example, to go back to the second phase of the policy life cycle, is that in the formulation of policy goals there is evidence of a lack of attention to issues of management once the desired policy shifts have been achieved. The focus of the actors is more oriented towards initiating and developing new projects instead of conserving and protecting existing values. According to the interviewees simple measures, such as a more strict observance of the zoning schemes, will provide major opportunities for closing the gap between the desired and the actual contribution. This is seen as the biggest task for the future.

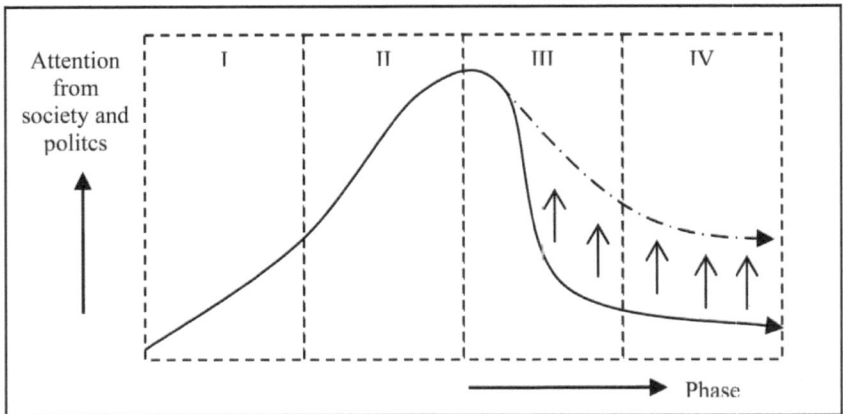

Figure 10.3   The policy life cycle and the attention from society and politics (derived from Winsemius 1986)

10.8   The potential contribution of actors

The previous section, in comparing the desired contribution with the actual contribution, showed that the gap is widening as the policy life cycle advances. However the conformity between the desired and the actual contribution in the early phases of the policy life cycle – especially in the phase of recognition – must be seen as a strong point in approaching the future. The parallel between the cultural heritage policy and earlier policy developments in environmental planning and water management shows that in order to reach the desired goals (i.e. that cultural heritage is seen as a substantial part of the quality of life) we must ensure that attention from society and in particular from politicians remains high. This must be realised by embedding the theme of cultural heritage into organizational structures, in the availability of economic and other resources for the realisation of policy goals, the development of knowledge and information systems and in social engagement. So public and political attention to the cultural heritage theme must be maintained or increased, especially in the later phases of the policy life cycle, as illustrated in Figure 10.3.

10.9   Reflections on the research method

Every actor interviewed during the course of this research had a positive view on the sustainable management of the Dutch Wadden Sea Region. However by using

Figure 10.4   Scenery of the Wadden Sea Region
(Photographs courtesy of Adriaan Haartsen)

the actor-consulting model it was found that this positive view had more to do with contributions they *desired* to make, and not so much with the contributions they were *actually* making.

A 'traditional' approach to research, based on opinions regarding indicator trends would without doubt have revealed that all actors are positive and supportive of the maintenance of cultural heritage of the area. However, one might have inadvertently come to the conclusion that the 'present contribution' to the sustainable use of cultural heritage was in good shape, and that a bright future was in prospect.

Instead, the actor-consulting model identified the question of *how* people want to contribute. Looking at the present contribution, actors were all found to be acting differently, according to their own interpretation of the policy goals, and that implementation of the goals and coordination between the parties involved requires considerable attention. Consequently, actor consulting has proved to have an advantage over traditional research methods by coming to the swift conclusion that cultural heritage has a bright future, but that a greater effort has to be made in practice to fulfil this ideal.

With regard to the coupling of the actor-consulting model with the Policy Life Cycle, we can say that the Policy Life Cycle offered us the possibility to cast a more structured and critical glance at different aspects of decision-making. More specifically, it has given us the opportunity to conclude that in the later phases of the policy life cycle the gap between the desired and the actual contribution was widening, and that extra attention should be given at this stage to *implementing* the cultural heritage theme more effectively and sustainably, with the support of a more structured and more thoughtfully resourced approach to the development and implementation of the project goals.

References

De Boer, R., L. Jager (2001) 'Beleidsverkenning landschappelijke waarden en cultureel erfgoed Waddenzeeregio' [Policy Exploration of Cultural Values and Heritage of the Wadden Sea Region], Department of Planning and Environment, Faculty of Spatial Sciences, Groningen (NL).

OCW, VROM, LNV, V&W [Netherlands ministries on culture, planning, agriculture and traffic) (1999] 'Nota Belvedere: Beleidsnota over de relatie tussen cultuurhistorie en ruimtelijk inrichting' [Memorandum Belvedere: Policy Document about the Relation between Cultural Heritage and Spatial Planning], VNU Uitgeverij, The Hague (NL).

VROM [Ministry of Housing, Spatial Planning and Environment] (1999) 'Vierde Nota Ruimtelijke Ordening Extra' [Fourth Memorandum on Spatial Planning Extra], RPD, The Hague (NL).

VROM [Ministry of Housing, Spatial Planning and Environment] (2001) 'Vijfde nota ruimtelijke ordening: Ruimte maken, ruimte delen' [Fifth Memorandum on Spatial Planning: Making Spaces, sharing Spaces], VROM, The Hague (NL).

Winsemius, P. (1986), *Gast in eigen huis: beschouwingen over milieumanagement* [Guest in Our Own House: Observations of Environmental Management], Samsom H.D. Tjeenk Willink, Alphen aan den Rijn (NL).

Chapter 11

Policy Appraisal and Sustainable Development in the North East Region of England

Geoff Porter[1]

11.1 Using actor-consulting as a policy appraisal tool

In this chapter, we explore how actor consulting was used to examine the outcomes of spatial policy at the regional level in the North East Region of England, and at the local level in the City of Newcastle upon Tyne. Research was carried out to explore the views of a range of actors, focusing on the extent to which they perceived plan policy to be promoting sustainable development, by reference to their experience in a major development proposal.

The process of participation in the spatial planning system began to develop in the UK in the 1960s. The conflict between on the one hand the British culture of freeholder rights to develop land in private ownership, with on the other hand the post-World War II tendency towards nationalization of development rights resulted in the Skeffington (1968) recommendations for participation. This attempted to balance the opposing forces by introducing legislation to incorporate a statutory requirement for public consultation in the Plan Review process. Consequently, at the time of the research, the principle of consultation in planning was long established, but conceived largely within a highly centralised system (Tewdwr-Jones 2002).

At that time, the UK national planning system had moved beyond the principle of 'presumption in favour of development' (DETR 1992), and the policy guidance was now focussing on 'sustainable development' (DETR 1997 and 1999b). An emerging feature of the plan review process was the requirement to carry out formal Sustainability Appraisals. This was at the time regarded as good practice, rather than as a mandatory requirement, because the research predated the implementation of the Strategic Environmental Assessment Directive (European Commission 2001). The Regional Planning Bodies and the local planning authorities were therefore reviewing and updating their land use plans in the context of this policy environment.

1 Sustainable Cities Research Institute of Northumbria University, Newcastle upon Tyne, and the Department of Planning and Transportation, Newcastle City Council, carried out the research work as part of the SUSPLAN project, which was part funded by the European Commission Interreg IIC North Sea Region Programme (Interreg 2001).

The research offered an opportunity to use the actor consulting method to review the regional and local approaches to the location of housing, which stood out as one of the key issues in the regional sustainability debate. Actor consulting would therefore provide a means to assess the performance of the regional and local plans in delivering sustainable outcomes, and provide an opportunity to relate this information to the formal appraisal processes.

11.2   National and regional policy environment

It is first necessary to examine the UK national policy structures that at that time underpinned the move towards 'sustainable development'. The discourse is complex, so our brief examination of it here is necessarily an over-simplification of the picture. Nevertheless, some clear indications can be given of the institutional perspective on the sustainability debate.

European-level guidance

The EC Directive on Strategic Environmental Assessment (European Commission, 2001: Article 2b) refers to *environmental assessment*, rather than the process of *Sustainability Appraisal* advocated in the UK. The Directive can be thought of as providing a minimum standard for environmental appraisal (Kornov and Thissen 2000). This contrasts with Sustainability Appraisal, which is based on a wider spectrum of social, economic and environmental criteria.[2] It is consequently important to bear in mind the overlaps and differences between plan appraisal processes pursuing definitions of *environmental* sustainability and those adopting a wider 'three-legged' approach, taking in social and economic concerns as well.[3]

---

2   According to Lee (2002), the potential benefits of integrated assessment tools such as Sustainability Appraisal, relative to more specialized forms of assessment such as Strategic Environmental Assessment, include greater benefits to decision-makers and other stakeholders, who need to see the full range of impacts rather than a subset; and greater ability to capture indirect effects that result from linkages between economic, environmental and social impacts. On the other hand, it is clear that the process of sustainability appraisal is at risk of being captured by the interests of a single dominant stakeholder or interest group: 'Sustainability Appraisal can in fact marginalize the very environmental and social appraisals that it is supposed to bolster as a counterpoint to dominant financial and economic interests [...]. We believe that the environmental component of sustainability appraisal must be strengthened as a condition for its retention' (Royal Commission on Environmental Pollution 2002: p. 98).

3   UK legislation and guidance on appraisal methods continues to evolve, but the tension between Sustainability Appraisal and Strategic Environmental Assessment remains. In an attempt to reconcile the two approaches, *Sustainability Appraisal of Regional Spatial Strategies* and *Local Development Frameworks* (ODPM 2004: 1.1.4) maintains that: 'The requirement to carry out a Sustainability Appraisal and a Strategic Environmental Assessment are distinct. However, it is possible to satisfy both through a single appraisal process. This

National guidance

Guidance on plan appraisal (DETR 2000b, ODPM 2004) recommends a formal technical review of the draft plan. Appraisal of both regional and local plans should explore *options* for the implementation of the spatial strategy, and inform the choice between options by using *sustainable development objectives* as the criteria for appraisal. Policies can then be appraised by seeking to refine them through successive iterations, using *targets* as the basis for appraisal. *Indicators* are to be used for monitoring the effects of the policies (DETR 1999a). This process of Sustainability Appraisal is used iteratively to develop the balance between economic, social and environmental factors in the spatial plan. In particularly, four national guiding principles for sustainable development are identified (DETR 1999a). These comprise: social progress that recognises the needs of everyone; effective protection of the environment; prudent use of natural resources; and maintenance of high and stable levels of economic growth. There has been extensive debate as to the mutual compatibility of these four guiding principles, with most commentators pointing out the difficulties inherent in maintaining high levels of economic growth, whilst at the same time maintaining progress on the other three.

Regional Sustainable Development Frameworks

The national government requires each English region to prepare a Regional Sustainable Development Framework (DETR 1999b) based on the four national guiding principles. These are envisaged as high-level documents that set out a vision for sustainable development, and the Region's contribution to sustainable development at national level (DETR 2000a). The intention is that Frameworks will provide a broad context within which economic and spatial planning at both local and regional levels will be developed. In response to the requirement for Regional Sustainable Development Frameworks, the North East Assembly produced a framework of objectives, targets and indicators in consultation with regional actors (North East Assembly 2002). This provides the means to guide the appraisal of both regional and local plans.

Local plans

The scope of English local plans is more limited than spatial plans in many other European countries (See for example: Williams 1996). English local plans cover key topics such as housing, commercial development, open space, the natural environment, conservation of the built heritage, and certain aspects of transport. Unlike many European planning systems, however, they do not directly guide investment in, for

---

guidance is intended to ensure that Sustainability Appraisals meet the requirements of the SEA Directive, and it widens the Directive's approach to include social and economic as well as environmental issues'.

example, water, energy, social services, health or education although all will be taken into account to some extent. UK plans are *indicative*, rather than *prescriptive*, in that unlike some other European countries (e.g. Netherlands), the local plan only seeks to indicate areas where development is in principle allowable, as opposed to areas where development will specifically be implemented on the ground. Local government in the UK must prepare returns to central government on a wide range of performance indicators. Lists of indicators are abundant, offering scope for either greater clarity or greater confusion, depending on how work on indicators is tackled and integrated.

11.3   Research methods

Opportunities for the research

It can be seen that planners at both regional and local levels are required to address sustainability of spatial plans via a formal appraisal system based on setting objectives and targets, supported by the monitoring of a potentially complicated set of indicators. There is an implicit understanding that a process of public consultation will take place to underpin this process. Tools to qualify public opinion in the development of spatial policy, such as the 'actor-consulting method', therefore offer an opportunity to address this need. This research was not however linked directly to a formal plan appraisal, but was regarded by the planners at Newcastle City Council as an opportunity to explore new participatory techniques for this purpose.

Selection of the spatial policy theme

'Location and Design of New Housing' was selected as the object for the research, because of its central role in the main strategic initiatives taking shape in the City at that time (see Table 11.1), and because of its ongoing role at the centre of the regional discourse on sustainable development (North East Assembly 2002).

Table 11.1   Links between case study theme and Newcastle City corporate strategies

Newcastle City Strategy	Strategic aims relevant to the Case Study
'Going for Growth'	Step-change in quality of housing within the City
Local Transport Plan	Encourage reduced distances travelled home to work
Biodiversity Action Plan	Increased numbers in key native species
Energy Strategy	Efficient energy design of housing in the City

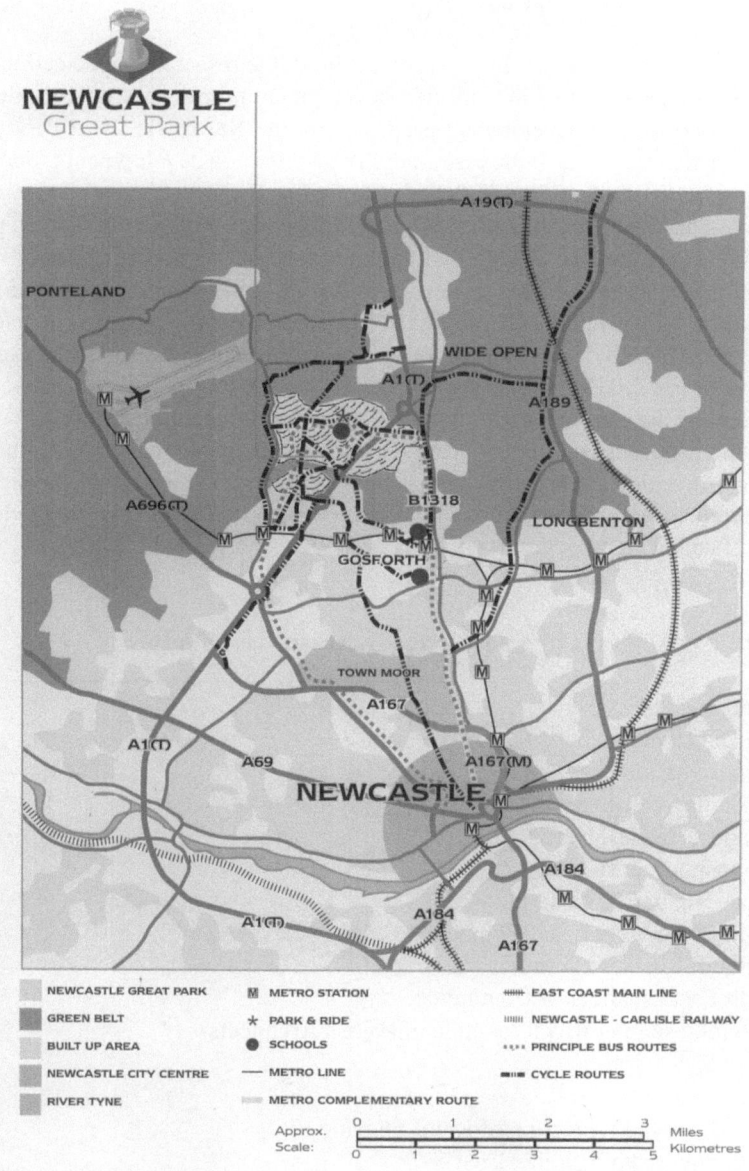

Figure 11.1   Diagram showing the location of Newcastle Great Park
Source: Bryant Group and Leech Homes 2000

Selection of the planning scenario

Rather than exploring policy in the abstract sense, the researchers decided to focus on the experience of actors in a specific development proposal. After considering a variety of options, the development proposals for the Newcastle Great Park (NGP) were selected as the focus of the case study. The NGP was a development proposal for 2,500 new houses and a technology park in an urban fringe setting, with easy access to road and air transport links (Figure 11.1). The land had originally formed part of Newcastle's greenbelt, so the land-use and transport aspects in particular of the NGP planning application had been controversial, with both strong support and strong opposition to the development proposals. The idea of developing the site had been pursued for a period of over ten years, and at the time of the research had reached the stage where final proposals for the design were being negotiated between the developer and the planning authority. This was a development, therefore, of regional significance with respect to the debate on sustainable development. By choosing to examine a development proposal of regional significance in terms of its physical size and strategic location, actor consulting offered the possibility of examining ideas for more sustainable approaches to policy at both local and regional levels.

Table 11.2   Actors taking part in the Newcastle case study

Business interests
   The Developer's consortium:
      House builders
      Planning consultants
      Landscape designers
      Transport consultants
   House Builders' Federation
   Regional Economic Development Agency
   Northern Business Forum
   North East Chamber of Commerce
   (also representing Royal Institute of British Architects)

Environmental interests
   Council for the Protection of Rural England
   Northumberland Wildlife Trust
   Countryside Agency
   English Nature
   Environment Agency
   Renew North (Regional renewable energy promotion)

Planning and Transport interests
   Government Office North East (regional office of national government)

Tynebikes (cycling organisation)

Community interests
Newcastle Healthy City Project
National Housing Federation (represented by a social housing company)

Selecting the actors and the scope of the discussions

The actors were selected to represent a broad range of social, environmental and economic interests in the NGP. The list of actors is shown in Table 11.2. The 'Sustainability Checklist' system (BRE 2002) was used to define the scope of the discussions with the actors. The Checklist provides a tool to evaluate the 'sustainability' of development proposals, and is intended for use by both planners and developers. It breaks down sustainability into eight categories: land use, transport, energy, natural resources, buildings, ecology, community and business, and steers the user through a set of criteria within each of these categories, as a means to appraise the development proposal. Despite its technical-rational origins (including the option for example for the user to apply a points-scoring system to the object of the appraisal) the Checklist provided a convenient tool to explore the actors' present, desired and potential contributions to the NGP design proposals.

Interview methods

The eight categories from the Checklist (BRE 2002) were used to structure the discussions, although only those that reflected the knowledge and interests of each actor were pursued. So for example those with specialist knowledge contributed information to only one or two of the categories, whereas others contributed to all eight. The actors were asked to provide a 'corporate' view of sustainability issues, and not a personal view, so although personal opinions were discussed from time to time, these were not formally recorded. This was effective in facilitating a clear picture of the actors' motivation, whilst respecting the sensitivity of the case study material. The present contribution towards sustainability included questions such as: What is the current behaviour of the actor? What is the current behaviour of other actors? What actually happened? The desired contribution included questions such as: How does the actor wish to behave, and what is preventing him/her from doing this? How could other actors contribute to achieving improved 'sustainability'? The potential contribution included questions such as: How might the gap between the present and desired situations be bridged? What new policies and approaches might be possible in order to improve sustainability? Are more sustainable approaches possible without recourse to formal policy? How can the planning system or the institutions that support it be improved to accommodate more sustainable outcomes?

11.4   Description of the analysis and evaluation

This brings us to the analysis of the actor interviews. The actors were able to contribute a wealth of information about how the planning application for the NGP development had proceeded, and how policy at national, regional and local levels had supported, or failed to support, their notions of 'sustainable development'. The actors were also able to contribute a variety of speculative ideas, which linked their perception of the gap between the present and desired states with the sustainability objectives of the North East Region (North East Assembly 2002).

Each interview summary was scrutinised to identify discreet sustainability issues, and the results were set up on a database. A selection of the records from the database is shown at Table 11.3. Each entry in the 'Potential Regulatory Ideas' column takes the form of an idea to improve sustainable development. These seek to link the gap between the Present Contribution and the Desired Contribution to one or more of the regional sustainability objectives (North East Assembly 2002), which are indicated in the column on the right hand side. Further columns in the database indicate the *nature* of the regulatory ideas (statutory, self-regulation, capacity building[4]), and the *level* of government that might be involved in their implementation (National, Regional, Local).

The analysis revealed that the contributions of the actors featured relatively little conflict. Similarly, there was little repetition of issues and ideas. It was decided therefore to adopt the issues identified as a checklist against which emerging spatial policy might eventually be evaluated. This database/ checklist therefore represents the output of the research.

The interviews not only revealed a high level of knowledge of sustainability issues, but also an unexpected desire to achieve a common purpose in this respect. For example, although house builders might be superficially thought of as lacking concern over environmental welfare, the interviews showed a strong recognition on their part that issues concerning transport, ecology, energy, building materials and community needed to be optimised to deliver a development that is perceived to be successful, and therefore more likely to be profitable.

The process of learning experienced by the planners who took part in the interview process developed their understanding of how the actors perceived the various sustainability issues, and consequently how policy might be structured to take into account the aspirations of the actors. Learning was however limited largely to the planners, because the results of the data analysis were not fed back to the actors. This perhaps emphasises the point (Chapter 8) that actor consulting should be regarded as an ongoing process, with review and update of the information passing

---

4   Note that fiscal regulation is generally not possible in a UK local policy context. Opportunities for 'capacity building' were instead considered to be of interest. Sustainable development capacity-building initiatives can be defined as 'all measures that strengthen governmental structures to meet the demands of sustainable development, as well as measures that create these capacities in co-operation with civil society' (Evans et al. 2005: 26).

Table 11.3 Extract from the results database, showing a selection of potential regulatory ideas generated from the gap between Present and Desired Contributions of actors; level of regulation (local, regional, national), type of regulation (direct, self, capacity building); and how the regulation might contribute towards achieving one or more of the regional objectives for sustainability

Actor no.	Category	Present Contribution	Desired Contribution	Potential Regulatory Ideas	Local	Regional	National	Type	Regional Objective
9	Community	Too many houses are being built in areas of low employment, perpetuating unsustainable living patterns	Smaller numbers of high quality homes should replace larger numbers of poor quality homes in areas of low employment	Regional plan accepts that a major re-adjustment of housing numbers is needed, as an aspect of achieving economic growth		Yes		Direct	Economic growth; Improve health; Ensure accessibility
8	Transport	Strategic location of NGP promotes road transport; proposed bus links to airport and city centre are inadequate	Tram or light rail network along riverside to West Newcastle and the NGP is needed, but public awareness and support is lacking	'Community Planning' process offers opportunity for local people to debate and influence local transport issues via the Local Strategic Partnerships	Yes			Capacity building	Participation in preparing the local plan
3	Land use	Citizens have a disadvantage relative to developers, who possess knowledge of the planning system, and finance to drive planning applications	Planning system must accommodate more effective input from citizens and elected representatives	Training in planning processes should be available to citizens. Elected representatives should be required to attend training in planning processes	Yes	Yes		Capacity building	Public involvement in decision making
6	Business	A major business development has been granted planning permission in the urban fringe, due to pressure on the City by businesses to release the land	Business development should take place on brownfield land (at transport nodes), but businesses regard such an option as insufficiently sexy	Can the major business networks be persuaded (e.g. Chamber of Commerce) that corporate social responsibility should extend to the use of recycled / brownfield land?		Yes		Self	To use non-renewable resources carefully

both to and from the actors on a regular basis. Only by this means will it be possible for the various frames of reference to converge, as mutual learning progresses.

11.5   Summary of the results

This section provides an outline summary of the database of actor contributions, based on the eight categories of analysis (BRE 2002). The relationship between the potential regulatory mechanisms and present and desired contributions to sustainability can be clearly recognised:

Land use

The discussions with the actors at Newcastle Great Park indicate that housing location is a regional issue, rather than a local one. This reflects the need to co-ordinate land-use planning for housing between local authorities to avoid displacement effects and to reduce greenfield land use. Linked to this is the desire for the resourcing of a regional team of spatial planners, one of whose responsibilities should be to develop a public communication programme covering alternative spatial strategies. New media to promote public and political debate might include computer modelling of land use for housing particularly at the local level, so that people can see existing land-use patterns and strategic options for the future. There is a need for more housing for single people and small families, and also a need for executive housing with easy access to city centres. As in the Groningen-Assen case study (see Chapter 14) there is a need to refocus on the types and sizes of homes being built, instead of the current preoccupation with targets for house building.

Transport

Many actors thought that instead of designing the houses with restricted parking facilities, the best regulatory approach would be to introduce parking restrictions at key destinations such as shopping malls or town-centre workplaces. The ideas for regulation therefore followed the view that as car ownership spreads, a coherent regional parking and access policy might regulate car use. This might help to alleviate the present situation, whereby Local Authorities must compete to attract, for example, retail shoppers arriving via car transport. A variety of actors suggested that road transport played too prominent a role in the NGP proposals. A number of regulatory and capacity-building ideas are proposed to make the decision-making process more transparent, and to steer the emphasis away from road transport.

Energy

There is a need to accelerate national-level fiscal support for domestic energy saving initiatives and renewable energy facilities. Without national fiscal support, regional

Figure 11.2   **Construction of housing at Newcastle Great Park, illustrating photovoltaic panel installation. PV is offered only as an optional extra for home buyers (present contribution). A desired contribution is for PV to be installed as a standard feature, but the financial aspects of this can be difficult to reconcile, despite UK government grants for domestic renewable energy installations (Photograph courtesy of Colin Percy, Newcastle City Council)**

and local initiatives will have little impact. For example, the house builders regard photovoltaic and solar hot water facilities as 'optional extras' for their clients, but despite government schemes to subsidise such installations, their uptake is rather limited (see Figure 11.2). Other fiscal opportunities include, for example, financial assistance to Housing Associations to take ongoing responsibility for home energy programmes, and support for domestic renewable energy 'net metering'. This latter point would make domestic production of electricity from renewable sources more attractive, by legislating for householders to supply the UK National Grid with electricity from sustainable sources at the same price as the purchase price from the Grid.

Buildings

The case study indicated a growing need for local and regional identity in housing *design*, which might be advanced by the publication of regional housing design guidance. There is a need to encourage *quality of workmanship* as the foundation for sustainable employment in the regional construction industry. Historically, the argument has centred upon the need for house building as a means to reduce regional unemployment.

Figure 11.3   Sustainable urban drainage has been a design feature at Newcastle Great Park, illustrating that the present contribution to urban water management in new development locations is now making considerable progress (Photograph courtesy of Colin Percy, Newcastle City Council)

Natural resources

The present contribution of actors such as the local planning authority and the Environment Agency towards urban water planning suggests that they are working more effectively together to deliver features such as Sustainable Urban Drainage Systems in new urban developments (see Figure 11.3). With regard to domestic construction however, it is doubtful whether self-regulation will be sufficient to encourage further uptake of, for example, grey-water systems and permeable hard surfacing. The house builders must be convinced of the commercial advantages of adopting these aspects of design. A stronger relationship between the house builders and the water companies might be helpful here.

Ecology

Further progress could be based on an existing outline scheme piloted by a neighbouring local authority to encourage the (self-regulatory) planting of native species in private gardens and on business premises. More local authority ecologists are needed (there is currently a major lack of resource) to agree planning conditions for new developments and to police the building sites to ensure that the planning agreements with developers are honoured.

Community

There was support for the notion that planning committees tend to take too simplistic a view of the social and environmental effects of development proposals, and that they tend to focus on the economic benefits. Training in land-use planning for Elected Members, with a focus on the complexity of the linkages between the economic, environmental and social impacts of building should be considered. It was suggested that housing growth in some geographical areas within the North East Region, especially the ex-coalfield areas, was unsustainable. Attempts at redevelopment and regeneration may actually perpetuate a sense of isolation from the rest of the world, and longer journeys to work. Here it may be necessary to accept the fundamental principle of locally managed decline in terms of *numbers* of households, whilst improving the *quality* of houses as an aspect of achieving a reduction in unemployment.

Business

Actors were generally supportive of the Newcastle 'clustering' strategy for business, including the principle that the NGP would help to provide a critical mass of related information technology businesses. It appears however that one of the main drivers for the release of greenbelt land at the NGP was due to pressure from businesses, who chose to link the concept of 'status' with a green field location for their offices. Although the building industry displays an increasing corporate awareness of sustainability issues, there is an opportunity for the promotion of the status of recycled land within the wider business community (self regulation).

11.6   Conclusions

The case study gave rise to a complex set of ideas for more sustainable policy and practice within the broad policy arena of 'location and design of new housing'. The wide range of issues identified suggests that a multiplicity of policies and activities must be introduced using a wide variety of instruments at local, regional and national levels, in order to improve sustainability (as seen from the perspective of the actors) in the policy area under investigation. The methodology was therefore effective in getting to grips with the complexity of the policy arena of housing location and design. Indeed, a key learning point for the planning officers and researchers taking part in this exercise has been to expose the sheer complexity of this policy arena.

The output of the research has centred on a list of issues against which emerging policy at local and regional levels can be evaluated. These issues are founded on ideas that are formulated by actors with direct experience of the shortfalls of contemporary planning policy in delivering sustainable solutions in practice. The weakness of this approach lies perhaps in the specificity of the NGP case. Whilst an exploration of a case study that features the *implementation* of policy might indeed produce an effective way of exposing sustainability issues, it also presents a weakness in terms of its *geographic* representativeness (Mason 2002). In order to address this point, it might be necessary for example to explore the issues identified so far with a more geographically representative selection of stakeholders. This might involve discussions with stakeholders on a wider geographic basis within the City (as a means of qualifying the results of the research at the local level), or alternatively on a wider geographic basis within the region (as a means of qualifying the results of the research at the regional level).

Neither has the research attempted to address the political dimension of policy development, by involving for example interviews with locally elected members of council, or elected members who represent regional interests. One of the inherent difficulties of the UK Sustainability Appraisal system is the conflict created by the potential incompatibility of the four national guiding principles for sustainability. The inclusion of economic growth here waters down the policy framework to what some[5] would see as an ineffectual compromise. For these reasons, the UK strategy can be categorised as a 'weak' (Pearce 1993) approach to sustainability. The Regional Sustainable Development Frameworks (DETR 2000b) will therefore set out objectives for economic growth alongside those for social progress, environmental protection and resource usage, and spatial policies will be judged against their ability to deliver on *all* these objectives. This will inevitably give rise to conflicts of interest in policy development, not least during the policy appraisal process. Scenarios of this nature require political decisions, despite the assumption of technical-rational theory that 'objective assessment' will lead straightforwardly to better decisions (Owens, Rayner and Bina 2004).

The checklist of issues that represents the output of this research may therefore help to take forward the processes of technical Sustainability Appraisal, but this will continue to be mired in the contradictions that underpin the UK national objectives for sustainable development. It can consequently be concluded that the process of learning on the part of the planning officers and researchers taking part in the actor consulting process has been a key outcome of this research, and that this might contribute in the fullness of time towards the development of better policy and practice at both local and regional levels.

References

BRE (2002) *A sustainability checklist for developments: a common framework for developers and local authorities*, April 2002, BRE, Watford (UK).

---

5   See for example the comments of the Royal Commission for Environmental Pollution (2002: 98).

Bryant Group and Leech Homes (2000) *Newcastle Great Park: Moving in the right direction, Bryant Group*, Newcastle Business Park, Newcastle (UK).

DETR (1992) *PPG1: Planning Policy Guidance Note 1: General policy and principles*, March 1992, The Stationery Office, Norwich (UK).

DETR (1997) *PPG1: Planning Policy Guidance Note 1: General policy and principles*, February 1997, The Stationery Office, Norwich (UK).

DETR (1999a) *Monitoring Progress (indicators for the strategy for sustainable development in the UK)*, July 1999, The Stationery Office, Norwich (UK).

DETR (1999b) *A Better Quality of Life – A Strategy for Sustainable Development in the UK*, May 1999, TSO, Norwich (UK).

DETR (2000a) *Guidance on Preparing Regional Sustainable Development Frameworks*, The Stationery Office, Norwich (UK).

DETR (2000b) *Good Practice Guide on Sustainability Appraisal of Regional Planning Guidance*, The Stationery Office, Norwich (UK).

European Commission (2001) 'Directive 2001/42/EC on the assessment of the effects of certain plans and programmes on the environment', *Official Journal of the EC*, 21.07.2001, Brussels (BE)

Evans, B., Joas, M., Sundback, S. and K. Theobald (2005) *Governing Sustainable Cities*, Earthscan, London.

Interreg (2001) *Interreg IIC North Sea Region: The Projects*, Programme Secretariat, Viborg (DN).

Kornov, L., and W.A. Thissen (2000) 'Rationality in decision- and policy-making: Implications for SEA', *Impact Assessment and Project Appraisal* Vol.18(3), pp.191–200.

Mason, J. (2002) *Qualitative Research* (2nd Edition), Sage, London.

North East Assembly (2002) *Sustaine – Quality of Life in the North East – Towards a Regional Framework*, January 2002, North East Assembly, Newcastle (UK).

ODPM, 2004, *Sustainability Appraisal of Regional Spatial Strategies and Local Development Frameworks*, September 2004, Office of the Deputy Prime Minister, London (UK)

Owens, S., T. Rayner and O. Bina (2004) 'New agendas for appraisal: reflections on theory, practice and research', *Environment and Planning A*, Vol. 36, pp. 1934–1959.

Pearce, D. (1993) *Blueprint 3: Measuring Sustainable Development*, Earthscan, London.

Royal Commission on Environmental Pollution (2002) *RCEP 23rd Report: Environmental Planning*, March 2002, HMSO, Norwich (UK).

Skeffington Committee (1968), *People and Planning: Report of the Committee on Public Participation in Planning*, HMSO, London.

Tewdwr-Jones, M. (2002) *The planning polity: planning, government and the policy process*, Routledge, London.

Williams, R.H. (1996) *European Union Spatial Policy and Planning*, PCP, London.

Chapter 12

Sustainable Urban Renewal: Opportunities for Drenthe Province

Dana Kamphorst[1]

12.1 Introduction

One of the main goals of urban renewal policy in the Netherlands is to contribute to the sustainability of the urban environment in physical, societal and economic terms. The need for a more integrated approach to urban development policy emerged at the end of the 1990s. This followed a series of urban developments that had failed to produce satisfactory social outcomes, despite the improvements that had been made to the quality of the physical living environment (TK 1997). The resulting urban renewal policy framework incorporates several policy fields at various levels of government, and relies upon concerted action by a variety of actors. The main focus for implementation of urban policy is however at the local level: municipal authorities must comply with urban renewal requirements in co-operation with the national government, and also the provincial authorities and local actors.

This chapter summarises a study, which sought to address the question of how the provincial authority of Drenthe, The Netherlands, might more effectively support the sustainable renewal of the urban environment in the Province. The research method was based on the actor-consulting model described in Chapter 8. The research explored the wishes, possibilities and constraints of actors with regard to the theme of 'sustainable urban renewal'. The picture that emerged from the discussions with the actors revealed that the notion of 'sustainable urban renewal' had a multiplicity of interpretations among the actors, and that the Province could play a role in facilitating a collective understanding of this fuzzy notion, via a variety of mechanisms.

12.2 Sustainable urban-renewal policy in the Netherlands

The aim of the Urban Renewal Act (November 2000) is to combat impoverishment in the cities through physical interventions. These interventions can be realised by co-ordinating cross-sectoral policy and combining financial resources at the local

1 Dana Kamphorst is researcher at the Department of Planning and Environment, Faculty of Spatial Sciences, University of Groningen, Groningen, the Netherlands.

level. One of the objectives of the new policy is to give municipal authorities greater powers and scope for policy-making. To this end, the government has made funding available in the form of an Urban Renewal Investment Budget (ISV), a combination of the subsidies formerly allocated to individual sectoral initiatives. The ISV funding allows the municipalities a relatively large amount of leeway in spending.

ISV funding policy makes a distinction between 'direct' and 'indirect' municipalities. The direct municipalities are the 30 municipalities having more then 100,000 inhabitants, which receive their funding directly from the government, while indirect municipalities are supervised by the provincial authorities that allocate the funding. The government's criterion for funding specifies that a municipality must draw up a long-term plan setting out its policy for the physical living environment (VROM 1999). The plan must provide a coherent explanation of the measures considered necessary to improve the quality of housing and the working environments, and to create a liveable physical environment.

The governmental framework for urban renewal policy states that a municipality must indicate in its development plan how interventions in urban renewal contribute to 'sustainable development' (VROM 1999). The definition of sustainability given by the government in the policy framework relates mainly to the physical environment. Consequently, the policies relating to 'sustainability' in municipal plans are mainly physical environmental tasks, such as noise remediation and soil decontamination. In addition, the municipality must pay attention to aspects such as sustainable house building, sustainable water management and energy conservation.

The policy framework of the Province of Drenthe also emphasises the need for sustainability. The provincial authority advocates a rather general definition of sustainability in order to allow the municipalities some flexibility in its interpretation. At the provincial level, sustainable urban development is defined as a mix of societal, physical and economic components. The provincial vision on sustainability is based on the philosophy that sustainability involves more than physical measures or measures relating to the environment.

The policy document on urban renewal (Drenthe Province 1999) contains an inventory of municipal plans for urban renewal in Drenthe. It shows that six municipalities in Drenthe have confirmed that they have the ambition to implement sustainable house building and a sustainable infrastructure. The inventory does not make this clear for the remaining six municipalities, so it would appear that they do not consider 'sustainability' a priority in urban development. 'This is remarkable because the government and the provincial authorities aim to stimulate and promote sustainable development in their current policy documents'(Drenthe Province 1999: 21).

This prompted the Province of Drenthe to conduct research, using the actor-consulting model, into the opportunities and constraints for promoting sustainable urban renewal in the Province. The actor consulting model was used to carry out research among several groups of actors and to establish how they would be willing to contribute to sustainable urban renewal, how they were already contributing to sustainable urban renewal, and how they thought they and other actors should contribute. The findings have been distilled into conclusions on the opportunities and constraints for the Province. The following section begins with a discussion of the relevant actors.

12.3 The actors

Urban renewal encompasses a wide range of objectives and activities, which should be implemented mainly at a local geographical level. In physical terms, this includes restructuring residential areas, regenerating industrial areas, improving accessibility and tackling urban environmental problems. The multitude of objectives relating to urban renewal requires the co-operation of many actors. The table below summarises their formal roles in the process of urban renewal.

Table 12.1   Actors taking part in the Drenthe Urban Renewal Study

Government	The central government is responsible for funding by means of the ISV system. The government creates a context for urban renewal in law and in the ISV policy framework, which specifies the requirements for municipal plans to address urban renewal. Within these frameworks, the government aims to create as much policy leeway as possible and to avoid a prescriptive role.
Province	'The Provincial Authority has sole responsibility for promoting and supporting urban renewal, particularly with regard to the municipalities that qualify for investment funding'(Bulletin of Acts, Orders and Decrees 2000: chap. 2, section 3). Thus the provincial authority, like the government, acts as a *budget provider*. The Province has a certain amount of leeway in allocating ISV funding, since it can apply its own criteria. For this purpose the Province also lays down a policy framework within the margins set by the national government. The Province also fulfils the role of *director*. In this capacity, the Province aims to support and encourage the municipalities (direct and indirect) within its territory. Furthermore, it aims to co-ordinate the various municipal urban-renewal policies at a supra-local level.
Municipality	Municipalities are responsible for the process of urban renewal, from the first sign of problems through decision-making to implementation and evaluation (Drenthe Province 1999: 6).

Other actors	The municipal authority co-operates with local actors and local residents to fulfil its urban-renewal obligations. The housing corporations are legally defined as structural partners of the municipalities.* Furthermore, residents, social organisations, local interest groups and local trade and industry are regularly involved in dialogue with the municipal authority. They are therefore involved in the planning process. Developers and investors, on the other hand, appear to be more incidental partners of the municipalities (SGBO/DHV 2000).

* Research by RIGO (2001) confirms the role of housing corporations as the principal partners of municipalities for area development within the framework of the ISV.

In this case study, the parties most directly involved in urban renewal in Drenthe were interviewed, namely actors from the Province, housing corporations and municipalities. The interviews also included, however, questions about how other actors, e.g. private individuals, could contribute. In the remainder of this chapter, statements from the interviews are presented in italics.

Figure 12.1   Urban renewal program 'Emmen revisited' by the municipality of Emmen (Photograph courtesy of Marcel Heemskerk)

12.4   Desired contributions to sustainable urban renewal

This section discusses the desired contributions from actors, namely, what they are *willing* to contribute to sustainable urban renewal. The municipal authorities, housing corporations, the provincial authority and residents will be discussed in turn.

Municipalities.

In general, municipal urban-renewal plans include the aim of *sustainable* urban renewal. The objectives are described using statements such as '*sustainability deserves our attention*'. However, sustainability is not high on the agenda in every municipality. This has to do with *'personal interests and the political climate'*. Generally, municipalities aim to create a pleasant social climate and pleasant 'sustainable' housing for their residents. For this purpose they aim to fulfil a directing role in the sense that they mobilise other actors: '*As a municipality we try to take the lead in the prevention of impoverishment. At the same time we try to involve other users and owners of the space in the entire project in such a way that, in the future too, the risk of impoverishment is limited because other actors are also concerned with effective spatial management*'.

Housing corporations

Housing corporations have an interest in the redevelopment of residential areas in the sense that '*if there are good housing conditions and a good living environment in residential areas, people have no reason to move*'. Generally, 'sustainability' is associated with '*the future value of the housing stock*'. If homes have *future value,* their occupants will not be inclined to move. When considering sustainability, housing corporations tend to see housing costs as the most important issue. This means that redevelopment must also benefit tenants and that investment in *sustainability* must yield definite returns in the form of cost savings. It can thus be said that housing corporations are in favour of *sustainability* provided that the cost increases will not be excessive. In principle, the housing corporations are only responsible for the housing stock and may therefore be willing to take measures that involve the use of energy-saving materials and flexible construction.

The provincial authority of Drenthe

As mentioned above, the provincial authorities use a rather broad definition of sustainability in order to allow the municipalities a certain amount of leeway in interpretation. The *Working Paper on sustainable development in the Province of Drenthe* states that 'the concept of sustainability refers to economic development that is in balance with what the neighbourhood has to offer. Sustainability should not be seen as a goal in itself. It is a societal process that evolves during a continuous and integral balancing of interests within the framework of the above definition'(Drenthe Province 2000: 7). 'In any case, sustainable development means not only a clean environment free of pollution and nuisance and resource conservation, but should also address economic and social aspects such as prosperity, the income gap, employment, social security, education, road safety, national-insurance contributions, working conditions, social participation, the living environment and international co-operation'(Drenthe Province 2000: 8). The Province sees its role primarily as that of '*director and process co-ordinator*'. The provincial director's role in the context of sustainable development can 'be fulfilled through examples and knowledge transfer (stimulating, supporting, co-ordinating and implementing policies)'(Drenthe Province 2000; 11). Furthermore, the Province sees its own role as stimulating other actors to consider *sustainable* development in the municipality in the context of their urban renewal responsibilities.

The public

Members of the public were not contacted directly in the case study. However, other actors emphasised the importance of considering public views and interests in the urban renewal process. Consensus building among residents for urban renewal and area redevelopments is considered essential for achieving sustainability. The needs of residents are one factor that influences the sustainability objectives of the housing corporations and municipal authorities. One aspect of this is cost. Actors from housing corporations and municipal authorities assume that residents will not usually be prepared to pay more for environmentally sustainable housing, whether newly built or renovated, or for a sustainable built environment. The need to minimise cost increases for residents is therefore a starting-point for actors when formulating an approach to sustainability. Another aspect relevant to public consensus is the fact that sustainable neighbourhoods or homes require structured management, and consensus among residents is considered essential for this.

This discussion on the desired contributions by actors reveals clear differences between them. Municipal authorities aim to fulfil a directing role, housing corporations aim to confine themselves to measures relating to the housing stock, and the provincial authorities seek to provide support and incentives. The following section will analyse the actual contributions by these actors. After that, we will be able to compare some of the differences between actual and desired contributions of actors, which will lead us to their potential roles in sustainable urban renewal.

12.5   Actual contributions to sustainable urban renewal

This section describes how actors have actually contributed to sustainable urban renewal. Again, we shall discuss the municipalities, housing corporations, the Province and the public, in that order.

Municipalities

'Municipalities are responsible for the process of urban renewal, from the first sign of bottlenecks through decision-making to implementation and evaluation'(Drenthe Province 1999b: 6). During the initial phase, extending roughly from problem definition to the drawing-up of a strategic plan for urban renewal, the municipalities formulate the aims and objectives for sustainable urban renewal. Most municipalities in Drenthe have already done this. At this stage, municipal actors regard it as a positive action that sustainability has been put on the political agenda: *'There is at least a policy statement that we are willing to do something'*. One municipal authority, however, mentioned that the urban renewal plan refers to sustainability *'only in a general sense. These ambitions must be elaborated before projects can be implemented'*. Translating general aims into actual goals to be put into practice in the Province of Drenthe is still *'at an exploratory stage'*, a *'research stage'* and a stage of *'consensus-building'*.

In the view of actors, research is needed to facilitate the translation of sustainability objectives into practical measures. Furthermore, it is necessary to gain experience in implementing sustainability measures. The municipalities still lack a clear and measurable policy on sustainability. Measures are not embedded into municipal policies. Several reasons have been given for this, for example: *'the environment department has a backlog of work'*, *'other actors are not co-operating'*, *'it is still a new subject'*, *'sustainability is not a political priority'*, or *'where money is involved, sustainability measures are scrapped in favour of a cheap and easy solution'*. A number of sustainability initiatives have been taken, however, at district and project level. There are examples of projects that increasingly focus on sustainability, often exploratory or pilot projects that are carried out on an *ad hoc* basis.

According to several municipal actors, their role in contributing to sustainability is a 'signposting' role. They cannot do much more than this because sustainability is generally regarded as 'something extra' to which no statutory requirements are attached. Currently, the role of the municipal environmental department is *'to follow developments during the process and identify the possibilities for sustainability'*. Several municipal actors indicate that the municipal environmental department should play a central role in sustainable urban development. On the other hand, it is important that sustainability becomes a permanent topic of discussion in the municipality: *'In fact, everyone should aim to do this in their own context. The main responsibility of the environmental department is to draw attention to sustainability'*. However, municipalities are restricted in their ability to deliver concrete results. Results depend on factors that are difficult to control. The majority of actors indicate that results mainly depend on *'the degree of interest shown by other people'*.

Housing corporations

For housing corporations, too, it is difficult to pin down any definition of sustainability in the context of urban renewal because this involves a wide range of measures that can be covered by the umbrella concept of sustainability. Various physical measures are mentioned in the redevelopment of existing residential areas: '*based on the concept of future value, two blocks of flats have been modified for senior citizens. The measures relate to 'use value' and saving energy*', '*an Agreement on Sustainable House Building for Drenthe has been implemented and, wherever possible, attempts are made to see its provisions as minimum requirements*', '*we have a model project for sustainable renovation*', '*in the context of sustainability, some houses are being demolished, others are being renovated and new houses are being built*' and '*houses are being insulated*'. The parties involved indicate that there is less scope for sustainability measures when redeveloping existing residential areas than for new housing developments.

The province

Most of the actors indicated that the role of the provincial authority is primarily a stimulating and co-ordinating role: 'The Province has few concrete opportunities to exert influence. The Province can co-ordinate, direct, influence and advise, but it has few definite tasks.' With regard to urban renewal, ISV subsidies have, however, increased that influence slightly 'When allocating funds, the Province can impose extra requirements with regard to sustainability'. The provincial policy for urban renewal is based on the prevailing Provincial Comprehensive Plan (Provinciaal Omgevingsplan: POP), which sets out a number of priorities relating to its urban renewal task. These have been elaborated in the Policy Document for the Stimulation of Urban Renewal (1999b). The priorities include the redevelopment of residential areas, the clean up of pollution-emitting factories and sustainable building and water management. These priorities have shaped the content of municipal urban-renewal plans and the project proposals drawn up by the municipal authorities.

The public

One example of public involvement in sustainable urban renewal is in the municipality of Emmen, which has implemented a project called Emmen Revisited. Local residents are involved in the plans that form the basis for the redevelopment of the project districts. Residents are also involved in the follow-up phase through neighbourhood teams and platforms that enable 'direct consultation with the neighbourhoods' (Projectbureau Emmen Revisited 2002: 1). Residents therefore feel that '*the local authority is listening to them*'. This is an important step in consensus building. Most of the interviewees described residents' wishes with regard to sustainability as: '*residents always wish for a safe and pleasant living environment*'. However, definitions of 'a safe and pleasant living environment' '*differ from neighbourhood to neighbourhood*'. This is precisely why it is essential '*to establish, for each urban renewal project, whether residents support sustainability*'. Actors from the provincial authorities, municipal authorities and housing corporations all hold the view that it is impossible to realise sustainability without public support.

The 'action' taken on sustainable urban renewal has so far failed to live up to the expectations of the public. Sustainability in urban renewal is still at an exploratory stage and measures are usually *ad hoc*. The view is that the municipal authorities lack a clearly embedded policy on sustainability. Moreover, the all-encompassing character of sustainability means that it is it difficult to control. In the following section, the potential contributions of all actors will be considered from the perspective of the shortcomings in establishing sustainability in urban renewal policy.

12.6   Towards potential contributions to sustainable urban renewal

To get to a point where all actors can fulfil their potential role in sustainable urban renewal, we need to know the discrepancy between desired and current contributions. 'Sustainability' is a broad and dynamic concept, the substance of which is defined through the collective interpretations of all actors. This is especially true for a theme such as urban renewal, where the concept of sustainability is still in an exploratory phase and requires detailed elaboration. Despite the emphasis on sustainable urban renewal in recent years, it is still not standard practice to include sustainability measures in projects for urban renewal and redevelopment. The transition from strategic policy goals to operational action-oriented objectives is still not complete. An essential point raised in the interviews was the fact that the actors, even those within the same organisation, define the concept of sustainability in different ways. These definitions are interwoven and form obstacles to a uniform vision of sustainability. From a theoretical viewpoint, this is a clear indication that a fuzzy planning issue requires clarification.

All actors recognise that sustainability requires an integrated approach. Just as it is necessary to strike a balance between economic, social and physical aspects, it is also essential that there is co-operation between the various policy fields. The interviews revealed that, although urban renewal obligations are being met on the basis of an integral approach, this is very difficult to achieve in practice. This is partly due to the unclear delegation of responsibilities. An example is the role of the municipal environmental departments, which, according to other actors, should seek to develop and communicate the latest thinking on sustainability. On the other hand, sustainability is a responsibility that must be shared by all the relevant policy fields. This may be prompting a 'wait and see' approach. Furthermore, when sustainability is discussed, municipal environmental departments do not always have an equal say in the process of balancing local interests because sustainability is regarded as 'something extra' and is very rarely given political priority.

To get towards a point where the various actors can fulfil their potential role, the first thing we can conclude is that the concept of sustainability needs to become clearer for each particular objective, project or proposed action. Furthermore, it is necessary that agreements should be made on the different roles that actors play, and that these agreements should have a more binding status than they do at present. Also, it appears that a knowledge gap still needs to be filled. According to the interviewees, there is a need for '*successful pilot projects for sustainable urban renewal*', '*technical know-how*', '*information exchange, for example through working groups on the theme of sustainability*' and '*support and stimulation through funding*'. These observations can be distilled into opportunities for the Province as a whole, which are discussed in the next section.

12.7   Recommendations for the Province of Drenthe

This study has shown, as far as the provincial authority of Drenthe is concerned, that the actual and desired contributions to sustainable urban renewal are reasonably in line. For example, the Province already aims to act as 'director', in a guiding and supportive sense. The Urban Renewal Act stipulates that the Province shall be concerned only with promoting and supporting urban renewal, in particular regarding municipalities that are eligible for provincial investment subsidies. The opportunities for the Province therefore lie primarily in continuing and developing its current role, especially since the other actors involved have indicated they still require guiding or directing support from the Province. The directing role of the Province incorporates a dual role, namely a role that relates to the 'substantive' component of sustainability (i.e. defining *what* must be done) and a role that related to the 'process' component (i.e. defining *how* can it be done).

With regard to the *substantive* component, we can conclude from the above that there is a need for a clear and concrete vision on sustainable renewal formulated by the Province itself, including a clear explanation of the level of *quality* that the provincial authority aims to achieve. This has to be done in close interaction with municipalities. The municipalities would then have a clear concept on which to build. The substantive opportunities therefore lie mainly in supporting the implementation of sustainable urban-renewal projects. With regard to the *process* aspect, the opportunities relate more to defining how sustainable urban renewal can be realised. According to actors from municipalities and housing corporations, the Province is in a position to steer the process of realising sustainability goals. In order to do this they must encourage co-operation between the relevant actors and policy fields. They must also support municipal environmental departments in creating a platform for sustainability. The Province has already been active in this field by generating enthusiasm and by providing inspiring examples and financial incentives.

The substantive and process components can be integrated by identifying policy themes relating to sustainable urban renewal that require more direct management and by identifying appropriate methods for doing this. Interviewees were therefore asked what form of regulation – by the provincial authority or otherwise – they considered necessary for a number of concrete policy themes relating to sustainability. Here, a distinction has been made between direct regulation (regulations issued by a higher level of government) and 'soft' indirect regulation (stimulus and support rather than prescriptive regulations).

Direct regulation for sustainability

With regard to sustainability in urban renewal, the interviews revealed a number of unambiguous issues that require direct regulation. Prescriptive measures imposed by a higher level of government provide a guarantee for municipal authorities that sustainability will be considered in the local process for striking a balance between various interests. This form of guidance is, then, primarily a source of support for local environment departments when balancing the interests of the various policies. The aspects of sustainability that actors regard as requiring prescriptive guidance are usually matters for which quantitative, measurable limiting values and target values already exist. These aspects relate mainly to physical themes (e.g. conservation measures, sustainable building) and supra-local co-ordination. The role of the provincial authority is set out in the POP (housing quotas) and current environmental legislation.

Indirect regulation for sustainability

Other concepts that interviewees relate to sustainability are more qualitative, for example: the quality of public space, efficient utilisation of space, liveability, quality of the living environment, preservation of cultural-historical values, sheltered housing, participation by residents in projects and public safety. The majority of interviewed actors indicated that these themes require elaboration at the local level. The reasons given were, for example, that these are context-related concepts that cannot be defined in a uniform way for all municipalities or districts and are difficult to quantify. These are themes that most clearly reflect the 'all-encompassing' character of sustainability and provide the greatest scope for the Province, particularly with regard to supporting the municipalities and other actors in elaborating and implementing these sustainability themes. Furthermore, local actors can work with the provincial authority in formulating and setting out targets for the implementation of sustainability measures. The opportunities for the Province, however, lie not so much in prescribing as in encouraging. With regard to processes, the Province could emphasise knowledge-sharing and information provision for local actors involved in implementing these qualitative and context-related themes.

Finally, it should be mentioned that different types of relationship exist between the provincial authority and the municipalities. This is very evident in the context of urban renewal. The nature of the relationship depends on the theme. On the one hand, the Province is considered to be a 'higher level of government' but, on the other hand, it is also regarded as a 'partner of the municipality'. These distinctions result in different roles for the Province. In the case of sustainable urban renewal, both relationships can be beneficial. As a 'higher level of government', the Province can fulfil a steering and directing role, in particular through the allocation of the Urban Renewal Investment Budget [ISV]. But it might fulfil a wider role as partner to the municipalities, a provider of resources and knowledge, a facilitator of knowledge-exchange and, where possible, a source of support for municipalities and other actors in the context of urban-renewal projects. This brings us to the following outline summary of the recommendations:

1. There is a need for a clear and concrete *vision* for 'sustainable urban renewal' in Drenthe, formulated by the Province itself. All actors would then have a clear concept on which to build.
2. The Province should develop its current role by providing inspiring examples of how to *implement* sustainable development, and via financial incentives, knowledge sharing and information exchange
3. In addition to this, considering the fact that the implementation of sustainability measures faces many difficulties, the Province can play a role in formulation and setting out targets for sustainability at local levels.
4. The Province can play a role in supervising the achievement of these targets.
5. The Province will have a limited role in direct regulation for sustainability; more opportunities lie in indirect regulation, the Province being a provider of resources and knowledge, a facilitator of knowledge-exchange and, where possible, a source of support for other actors.

12.8   Observations on the research

The main advantage of using the actor-consulting model to gain insight into sustainable urban renewal was the fact that it was possible to identify discrepancies between the actual and desired contributions of actors. The study showed few discrepancies in this regard between the actual and desired contributions of provincial authorities and the housing corporations. However, there were considerable discrepancies at the municipal level. The model enabled an inventory to be made of these problems, which appeared to be concentrated around particular themes. In the first place, there is little agreement on the question of how to conceptualise sustainability, and consequently the municipalities are encountering problems in translating strategic objectives into concrete goals for implementation. Secondly, the fuzzy character of sustainability means that there is a lack of clarity among environmental departments and other actors as to their precise role. The analysis of these problems has made it possible to identify areas of opportunity for the Province to develop sustainable urban renewal, mainly by adopting the role of facilitator in support of the key actors.

References

Drenthe Province (1999) *Beleidsnotitie stimulering stedelijke vernieuwing* [Policy Document on the Stimulation of Urban Renewal], Provincie Drenthe, Assen (NL).

Drenthe Province (2000) *Discussienota duurzame ontwikkeling provincie Drenthe* [Working Paper on sustainable development in de Province of Drenthe], Provincie Drenthe, Assen (NL).

Projectbureau Emmen Revisited (2001) *Jaaroverzicht* [Annual report], Gemeente Emmen (NL).

SGBO/DHV (2000) *Evaluatie ISV plannen. Evaluatie van de meerjaren ontwikkelingsprogramma's van de rechtstreekse gemeenten* [Evaluation of the Urban Renewal Investment Budget. Evaluation of the multi-annual development programmes of the direct municipalities], The Hague.

Staatsblad (Bulletin of Acts, Orders and Decrees) (2000) *Wet van 15 november 2000 ter stimulering van integrale stedelijke vernieuwing* (Wet stedelijke vernieuwing) [Act of 15 November 2000 for the promotion of integral urban renewal (Urban Renewal Act)], Nr.504, Sdu Uitgevers, The Hague.

TK (Tweede Kamer / House of Parliament) (1997) *Stedelijke vernieuwing. Brief van de staatssecretaris van Volkshuisvesting, Ruimtelijke Ordening en Milieubeheer* [Urban Renewal. Letter from the Netherlands Minister of Housing, Spatial Planning and Environment], Vergaderjaar 1996–1997, 25427, nr.1, Sdu Uitgevers, The Hague.

VROM (1999) (The Netherlands Ministry of Housing, Spatial Planning and Environment) *Beleidskader stedelijke vernieuwing* [Policy outline urban renewal], Information bulletin, Directie Voorlichting, The Hague.

Chapter 13

Sustainable Urban Water Management: Opportunities for the Province of Drenthe

Rob de Boer[1]

13.1 Introduction

Following on closely from the theme of the last chapter, this chapter describes a research project into policy approaches towards sustainable urban water management in the Province of Drenthe (see De Boer et al. 2002). The aim of the study was to explore the possibilities and limitations for the provincial authority in developing its policy for urban water management. According to the Provincial Comprehensive Plan for the Province of Drenthe (Drenthe Province 1998), 'sustainability' must be a priority in all aspects of provincial policy. As already discussed in previous chapters, 'sustainability' remains a fuzzy notion from the perspective of most of the actors involved in the development and implementation of planning policies in Drenthe. The purpose of this research was to disentangle the fuzzy nature of urban water management, which is set out as one of the policy sectors in the Drenthe provincial plan.

This chapter begins with an outline of urban water management in the Netherlands (§ 13.2), followed by a survey of the complexity of the administrative structures (§ 13.3). We then present our analysis of the situation in Drenthe, based on interviews held with actors. Sections 13.4 and 13.5 describe, respectively, the actual and desired contribution to sustainability of the actors involved. Section 13.6 describes the recommendations of the study, in terms of the Province's capacity for contributing to sustainable urban water management. Finally, Section 13.7 offers some observations on the actor-consulting model, seen from the perspective of the study described in this chapter.

13.2 Urban water management and sustainability

Ideas about urban water management in the Netherlands are changing. Owing in part to the country's geographical situation and low-lying position, the historical aim of water management has always been to drain off surplus water as fast as possible. In the past, small meandering rivers were straightened, streams were deepened and the entire country was covered with a network of ditches, channels and other

---

1   Rob de Boer is researcher at the Department of Planning and Environment, Faculty of Spatial Sciences at the University of Groningen, Groningen, The Netherlands.

infrastructure, all designed to increase drainage capacity and keep Dutch feet dry. Water-management methods were mainly technical and there was great faith in technology. However, in recent years, the onset of more extreme climatic conditions has shown us that this approach carries large risks. There is a growing awareness that technological advances and know-how are not enough to solve the problems relating to water management. Following such a technocratic approach can lead to increased safety risk and economic damage as a result of flooding, and to the increasing loss of ecological, landscape-related and environmental qualities. It is also becoming clear that there is a strong correlation between issues of water management and other policy issues relating to the physical environment (e.g. environmental policy and spatial-planning policy). These developments mean that we need a new approach to water management. The concept of integrated water management was introduced in the Netherlands progressively between 1985 and 1998 (V&W 1985, 1990 and 1998). Recent events, such as flooding rivers and the continual threat of rising sea levels, have only accelerated this process.

In addition to the growing focus on water management in general, there is a particular interest in water management for the *urban* environment. Urban water management refers to matters such as the open water systems of interconnected canals, rivers and ponds, the quality of drinking water, separation of rainwater and wastewater, groundwater, the attractiveness of living near open water, the recreational value of water, and wetland development. In this urban water management can be seen as an 'umbrella concept' that covers a variety of activities. The addition of the concept of *sustainability* to urban water management makes matters even more complicated. In accordance with the broadly accepted definition of sustainability as formulated by the Brundtland Committee, 'opting for a sustainable water system means anticipating future developments and developing a policy in which great importance is attached to long-term benefits. Such a policy also entails that no decisions are taken now that make it impossible for society in the future to respond to changing conditions. We should only take measures that we will not regret later' (Dutch Commission on Water Management in the 21st Century 2000: 36). However, by means of the actor-consulting model we deliberately choose not to start from a prescribed definition of 'sustainability'. Rather, actors should be consulted regarding their views on sustainability in general and in this case on sustainable urban water management in particular. Therefore, first we will discuss the actors involved in urban water management.

13.3 The actors

In the course of the 1990s, urban water management has become a relevant theme for the various authorities. As mentioned above, it is a relatively new theme for which policy consequences are hard to predict. The good intentions concerning urban water management have not yet been consolidated into agreed activities or procedures and there is only a limited idea of their possible results. Equally, it is often unclear how actors can work together and share responsibility for policy-making.

Toonen (2000) characterizes water management as a layered and multifaceted policy arena in which a variety of actors and administrative organizations play a role. The setting is a polycentric one in which a variety of interests, organizations and institutions in the public and private sectors either have or demand a share. Water management is a typical responsibility of elected government or, as Toonen (2000: 15) puts it, of 'constitutional interest'. Protection from flooding, the development of sustainable water systems and methods for distributing a limited amount of water among users cannot be left to private initiatives. The central government, local authorities, district water boards, property owners and water companies each have their own role within this 'constitutional interest' (compiled from Van den Heuvel et al. 1996: 88, and the Rathenau Institute, 2000: 30).

Table 13.1   Responsibilities of the main actors in urban water management

Central government	Strategic planning, coordination and regulation (central government in general)
	Policy on water management and the treatment of waste and sewerage water (Ministry of Transport, Public Works and Water Management)
	Policy on drinking water and sewerage (Ministry of Ministry of Housing, Spatial Planning and Environmental Management)
	Policy on wetlands (Ministry of Agriculture, Nature Management and Fisheries)
Provinces	Strategic planning, coordination and regulation
	Issuing permits for groundwater extraction and filtration
Municipalities	Implementation of sewerage policy and urban drainage in general
	Decisions on sewer construction and renovation
District water boards	Regional water-system management and treatment of wastewater and sewerage
	Water-level management and supply management in urban and rural areas
	Surface-water quality management
Property owners	Condition of buildings and private drainage systems
Water companies	Public water supply and thus groundwater management and the collection, production and distribution of drinking water

13.4   Step 1: The actual contribution to sustainable urban water management

Given the scale and nature of urban water management in the Province of Drenthe, the study was structured on actor categories defined under the headings of 'provinces', 'district water boards', 'municipalities' and 'the public' (including property owners). This section outlines the activities for each actor category, covering their current contribution to sustainable urban water management.

Municipalities

As stated in the discussion document *Gemeenten geven water een plaats* (translated as: 'Municipalities give Water a Place') (Association of Dutch Municipalities 2001), water should be a structural municipal responsibility, but this is not yet the case. In general, municipalities prefer to adhere to a rigid interpretation of current legislation. This is partly due to the mindset of staff. The municipal authorities often lack people who can 'think outside the box' in a truly coherent way. It is also difficult for a municipal authority to look beyond its own municipal boundaries, and by their very nature, issues of water management often transcend boundaries. Municipalities are mainly concerned with their own problems and housing quotas, and regional considerations are often overlooked. Regarding the current contribution to sustainable urban water management, the urge for expansion driven by economic factors sometimes stands in the way of the wider application of sustainable solutions. Yet municipalities *do* have ambitions with regard to sustainability. These ambitions mainly relate to measures such as the separation of waste water and rainwater drainage systems, making water visible in the urban landscape, separating rainwater and wastewater, and sustainable construction materials for roofing and gutters. Water management is not yet a determining factor in selecting potential locations for urban development, although it must be said that most existing development locations were selected at a time when water was a less prominent theme. It is hoped – and here we anticipate the desired contribution of municipalities to sustainable water management – that this will improve in the future. Urban renovation offers possibilities for improvement of the water system as it can be carried out in combination with other interventions (e.g. preparing construction sites and road management). However, without urban renovation, these improvements would not take place.

District water boards

The past few years have seen rapid developments in the policies pursued by district water boards, partly owing to the recent focus on water management in the Netherlands. District water boards often know more about water and water systems than the provincial authorities, certainly as a result of their recent expansion. In some cases, however, the water boards lack the means to apply their expertise. Their current contribution to sustainability mainly involves assigning functions based on the hydrological position in the catchment area, raising public awareness of water (e.g. by making wastewater more visible in urban areas) and, finally, ensuring that water is a determining factor in the selection of potential locations for urban development and planning.

Provinces

The Dutch provincial authorities – and here the reader should bear in mind that the Province of Drenthe has commissioned this research – recognise that maintaining a balance between water and the environment at the local level is largely the task of municipalities and district water boards. The Province of Drenthe is responsible for supervisory policy, reviewing the management plans of the district water boards and the land-use plans and sewerage plans of the municipalities. In cases of non-conformance by other parties, however, they rarely enforce corrective measures. The Province of Drenthe is currently contributing to 'sustainable' water management by adopting the approach of incorporating it into urban development in existing areas.

The public

At present, there is very little public participation in planning procedures, despite the fact that this is vital to consensus-building and thus to the success of planning solutions. In general, members of the public only take action when a decision appears to have a direct impact on their own situation, and not when it affects them indirectly or in the far future. The selection of potential locations for urban development is often a strategic decision in which very few people are directly involved. Furthermore, such decisions often remain confidential for a long time, partly to ward off speculative property developers. Nevertheless, public participation in planning procedures is increasing. For example, when people move to a new area, they often seek to become involved as new residents in the planning procedures of the neighbourhood to which they are moving. Finally, processes of urban renovation have the highest level of public involvement. However, this often concerns matters that are directly visible and yield direct benefits (e.g. safety and waterfront living) rather than less visible issues such as recycling grey water, or separating rainwater and wastewater.

The survey of actor activities shows that the arguments for focusing on urban water management in general and sustainable water management in particular are somewhat diverse. We will return to this point at the end of the next section.

13.5   Step 2: The desired contribution to sustainable water management

Municipalities

With regard to the actual contribution made by municipal authorities in the Netherlands, we have stated above that sustainable water management is hampered by economic motives and the inability to look beyond municipal boundaries. In the near future, municipalities would like in principle to address flooding, improve safety, and improve the sewer system in internal as well as external areas. They would also like to introduce more open spaces with water to improve the quality of the physical environment. An important development in municipal policy is the creation of a municipal water-management plan. This is expected to be a powerful instrument that can achieve clear results in practice. A number of district water boards intend to draw up a water plan with all the municipalities, and are offering advice and financial support in this.

District water boards

The district water boards' approach to urban water management has five main aims. First, water must be a guiding factor in environmental policy. The water boards intend to increase their influence on environmental plans and urban development in order to make water management more sustainable. In the second place, they intend to make a greater contribution to urban water management by contributing in financial terms and by providing planning expertise to municipal authorities and other actors (e.g. property developers). Third, they wish to shift the management focus from 'cleaning-up' to 'keeping clean'. Clean rainwater should not drain into the sewer system, but should be kept clean and drained away more slowly or recycled. The fourth aim is to improve communication. The success of urban water management depends partly on the information and support available to actors. In the future, more attention will be paid to communication with municipal authorities and the general public. The fifth aim is to increase the awareness that interventions in the water system involve a different timescale compared to other planning decisions. Changes in the water system are less easily to undo, and affect people for a longer period than other planning interventions. An example of such an ill-considered change is that of the large number of canals and ditches that were filled in some thirty or forty years ago, and the present plans to reopen them again.

With regard to their desired contribution to sustainability, the district water boards are looking to shift their focus towards – or rather *back* towards – more natural water-management methods and to implement specific measures for this purpose. By making water more visible as a natural feature in urban areas, the water boards hope to increase the public's awareness of water. Furthermore, when potential locations are selected for development, water-management aspects should also be considered. The history of the Netherlands is characterized by a struggle against water. Settlements were traditionally established on sand ridges, particularly in marshy Drenthe province. In the recent past, largely as a result of our faith in technology, development has taken place in locations that are relatively unfavourable from the perspective of water management.

Provinces

The provincial authorities in the Netherlands are playing an increasingly important role in realising sustainable water management. Their main role should be a directing role (Association of Dutch Municipalities 2001: 8, National Institute of Freshwater Management and Waste Water Treatment 2002: 16). The provincial authorities co-ordinate environmental planning, water management and conservation. Their directing role means, for example, that they should know where to obtain the relevant knowledge and how to deploy the skills of other parties (district water boards, municipalities, waterworks, advisers, etc.) to achieve shared goals. The directing role of the Province is primarily to facilitate the process of finding creative solutions by making the best possible use of the powers and expertise of actors. This involves, for example, providing information on ideas about sustainable development, promoting an integral approach and implementing methods to improve decision-making. The Province is prepared to assume a directing role, particularly in the case of cross-sectoral problems. This is reflected in the plans for water catchment areas.

The public

The participation of private individuals (members of the public and property developers) is important in terms of co-financing and raising awareness of the need for sustainable water management outside the limited sphere of the local authorities. According to the interviewees, the desired contribution to sustainable water management of members of the public is largely determined by two factors: the awareness of water and the price that we – as a society and as individuals – are prepared to pay for it. Everyone employed in the Dutch water sector agrees that water management should be able to respond to both the current situation and to our expectations for the future. But the public is not yet sufficiently aware of this need. Communication and awareness are confined to the 'inner sanctum' of authorities and an extremely small section of society. The wishes of individual citizens depend strongly on the extent to which they are or become aware of the need for sustainable water management. Here, we come to the second point: the price of urban water management. How far do we want to go in adapting urban water management in general and sustainable solutions in particular? The interviews show that, on the whole, the wishes of the public relate mainly to tangible matters such as safety, flood prevention, clean and clear surface water and proximity to open water as a feature of an attractive physical environment (water for recreation and waterfront living).

The views of the actors on sustainable urban water management diverge to an extent, but also show a certain cohesion. The differences between actors can largely be explained by their arguments in favour of sustainable urban water management. Municipal authorities are keen to promote their own economic interests. District water boards are concerned with the quantity and quality of groundwater and surface water and, finally, many private individuals focus on their interests as property owners. The cohesion is explained by the common desire to recognise the value of urban water management as a policy theme. Based on the differences and similarities in the actual and desired contributions to sustainability, we have drawn a number of conclusions and made recommendations for the Province of Drenthe, described in the study as 'opportunities'.

13.6   Conclusions of the study and recommendations for the Province of Drenthe

This section focuses on the opportunities for sustainable urban water management in the Province of Drenthe. As mentioned in section 13.1, the main goal of the study described in this chapter was to investigate where the opportunities lie for provincial authorities. Here, a meaningful distinction can be drawn between selecting potential locations for urban development, planning the new urban development, and finally urban renovation.

Selecting potential locations for urban development

Selecting potential locations relates to the search for new urban development areas. The process of selection is partly based on the features of the local water system. Here, we emphasise partly. Various policy sectors and various levels of administration set preconditions that determine the decision-making process. These include the existing pattern of urbanization, the presence of nature reserves, existing infrastructure, administrative boundaries etc. In practice, the selection of potential locations makes virtually no allowance for the features of the local water system, although water should be one of the main determining factors. The selection of potential locations is very important. Once development has taken place, there are, from the point of view of the public, only a limited number of ways to compensate for inappropriate selections. It is important too that selected locations meet the criteria relating to irreversible effects on water systems. It is also vital to take a long-term view when assessing the necessary management measures. This is the only way to guarantee safety and prevent flooding in the long term. We have already seen that, from a hydrological perspective, certain municipal decisions have proved unfortunate. According to the interviewees in the study, the provincial authorities can fulfil an important role as 'a directing authority'. The provincial authority, more than any other, has a supra-local view of desirable developments (see Chapter 4). In the first place this involves decisions regarding trans-boundary catchment areas and long-term choices. Here, municipalities should make more use of the wider spatial-planning context of 'the Province'. The provincial authority could stipulate, for example, 'no urban development in stream valleys or heavily drained polders, unless conditions or counterarguments can be indicated'.

It is not easy to establish 'water' as an element in spatial planning. The provincial authorities' criterion for future spatial development is the 'water assessment', which helps to determine whether a proposed land use is suitable or unsuitable from the perspective of water management. The 'water assessment' process is used in all spatially relevant decisions by central government, provinces and municipalities in national and local water management. Given that spatial planning and water management are regional responsibilities, the provinces are an important actor in the success of water assessment. In most cases, when water assessments are carried out the provincial authority will act as plan reviewer (except when the province itself is the initiator, for example in the reallocation of land use and larger infrastructure plans). The plan reviewer approves or rejects an environmental plan/decision in part or in full. This is done in accordance with the relevant policy framework and the advice provided by water managers in the context of the assessment. Here, the role of the province is mainly a guiding one, aimed at stimulating cooperation between the authorities involved. Further, the Province of Drenthe can, in a more emphatic way than is currently the case, carry out water assessments when approving location decisions. Criteria for the selection of new urban locations include the interrelation of spatial functions, buffer zones for the separation of environmentally harmful and environmentally sensitive functions, the management of floodwater and pollution and, crucially, creating enough space for water (in relation to pressure on areas for drinking-water supply and allocating retention and overflow areas).

Planning new urban development

In general, the development of new urban areas is regarded as more favourable for realizing sustainable urban water management than renovations in existing urban areas. New urban areas offer completely new opportunities for urban planning, such as natural water drainage, retention, purification and re-use of the area's own water and hydrological cycles. However, when an existing area is renovated, the effectiveness of the planning process is heavily influenced by choices made in the past. Therefore, the most important argument in favour of new urban developments is that they enable planners to start with a clean slate. The provincial authority will thus fulfil a different role in a new area than in an area to be renovated, where municipalities and district water boards also have responsibilities.

Every actor involved in the planning and development of a new urban area has his own experience of water management. The town planner, for example, can include water as a structuring or cultural-historical element in his plan, while the main concern of property developers is surface water, which allows them to realise waterfront developments. Nature organizations will be interested in the extent to which water contributes to ecological qualities, such as biodiversity and wildlife corridors.

When new urban areas are developed, the municipal authority usually takes the leading role, sometimes in co-operation with a private investor (business, housing corporation). The purpose of a municipal master plan or land-use plan is to make clear spatial-planning choices determining, for example, how land-use will be allocated in the future and how a balance will be achieved between the interests involved. The development of new urban areas requires interventions in the physical structure and it is often necessary to build extra infrastructure or adapt existing infrastructure. In most cases, the municipalities and district water boards will build the necessary infrastructure. In this way the district water board can meet the requirements of the water system, while anticipating future needs and taking account of sustainability with regard to water in the environment. It goes without saying that the district water board should therefore be involved in the planning process from the outset (site selection, master plans). In this context, the role of the provincial authority is limited. Moreover, the interviewees see no immediate need to change this situation. However, if we look at plans for 'sustainable' water systems in new development areas, we see that they are often highly activity-led and lack an underlying structure. Almost every case lacks a vision of what 'sustainability' means for the plan or for the management of the water system. A role for the Province here – together with the district water board – could be to stimulate an integrated vision of 'sustainability' by means of information (indirect communicative control), implementation subsidies (specific economic control) and an improved review procedure for structural concepts and land-use plans (more compulsory forms of control). A further point to bear in mind is that, although municipalities promote integration and sustainability at the beginning of a planning process, a number of ambitions fall by the wayside as implementation approaches. When planning a new urban development, authorities can set fairly ambitious goals for water-neutral or even water-positive development. The authorities should encourage each other to do this. Thus, the Province of Drenthe is in a position not only to include 'water' as a compulsory policy sector in land-use plans, but also to specify which issues should be addressed.

Urban renovation

For an existing urban area, the situation is usually more complex than for a new area. An important difference between planning new development and urban renovation is the presence of residents, existing infrastructure and history. Land uses have already been allocated. These elements place constraints on urban renovation. On the other hand, renovation also offers plenty of opportunity for improvement. Here, ambitions depend on the feasibility and value of sustainable water-management measures in relation to the totality of measures.

In general, existing cityscapes have usually been shaped by cars and housing. In post-war neighbourhoods for example (see Chapter 14), water is seen as an open space. When these neighbourhoods were built, the priorities were drainage and safety, and all such areas have a 'combined' sewerage system. Because wishes change over time, many towns and cities are co-ordinating plans for neighbourhood renovation with activities that are already planned – partly to keep costs to a minimum. The feasibility of a new housing development will depend mainly on its distance from a watercourse into which water can be discharged. Feasibility is also determined by the permeability of the soil. In existing urban areas, the feasibility of constructing sustainable water systems depends on demolition activity, road reconstruction and sewer renovation. The focal points of urban renovation are space for extra water and facilities such as water-friendly construction and renovation, or the modification of drainage systems that can act as reservoirs and/or filtration facilities.

Opportunities for the Province of Drenthe. To summarise the above: the opportunities for the Province of Drenthe lie mainly in selecting potential locations for new urban developments and renovating existing urban areas. There are fewer opportunities with regard to new urban developments. At first sight this appears illogical, but the explanation is extremely simple. Owing to initiatives taken by municipalities, district water boards and other actors, sustainable water management is already a relatively important theme in new urban-development projects. However, the 'added value' created by a supporting and/or co-financing Province is outweighed by the 'sustainable added value' that an active Province can create – using the same resources – by adopting a 'sustainable' approach to selecting potential locations and renovating existing urban areas. This does not mean that provincial authorities should take a back seat in the development of new urban areas, rather that provincial policies should focus on introducing a 'sustainable' approach (for example via a water assessment) to selecting potential locations and renovating existing urban areas. In spite of all the attention that sustainable urban water management has received in recent years and the consensus regarding its 'added value', results 'on the ground' are rather disappointing. A number of successes have been achieved, but the transition is by no means complete. A considerable period of time may pass before structural, sustainable urban water management becomes the rule rather than the exception. In the meantime, many opportunities created by new developments and activities in existing neighbourhoods may be missed. It is partly the responsibility of the provincial authorities to ensure that the number of missed opportunities is kept to a minimum.

13.7 An observation on the actor-consulting model

In this study, the main value of the actor-consulting model has been to illustrate policy approaches to urban water management in the Province of Drenthe. In theory the responsibilities are equally divided among the actors. In practice, however, the situation is different. The provincial authorities must redress the balance and fill the 'gaps' – referred to in this study as 'opportunities' – or indirectly stimulate other actors to do so. The actor-consulting model has proved to be a valuable tool for detecting shortcomings in policy approaches to urban water management in Drenthe, precisely because it is not based on fixed indicators but on the everyday experiences of actors. By consulting these actors we could draw conclusions and make recommendations, which clarified the fuzziness of sustainability in urban water management. The results of the research will be input into the policy framework of the next Provincial Environmental Plan. There is no doubt, that by consulting the actors involved, the external support for the new policies will be broadened, and that the new policies will be a step towards a more sustainable Drenthe.

References

Association of Dutch Municipalities (2001) *Gemeenten geven water een plaats: Discussienota over de positie van gemeenten op watergebied* [Municipalities give Water a Place. Discussion Document on the Position of Municipalities in the Water Management Arena], Association of Dutch Municipalities, The Hague.

Boer, R. de, Kamphorst, D.A., Roo de, G. and M.J.C. Schwartz (2002) *Kansen voor duurzaamheid bij stedelijk waterbeheer en stedelijke vernieuwing: de provincie aan zet* [Opportunities for Sustainability in Urban Water Management and Urban Renewal], Faculty of Spatial Sciences, University of Groningen, Groningen (NL).

Commission on Water Management in the 21st Century (2000) *Waterbeleid voor de 21e eeuw: Geef water de ruimte en de aandacht die het verdient: Advies van de Commissie Waterbeheer 21e eeuw* [Water Policy for the 21st Century: Give Water the Space and the Attention it Deserves], Commission on Water Management in the 21st Century, The Hague.

Heuvel, H. van den, Jacobs, J.C.J., van Eck, P. and E.T. Schutte-Postma (1996) *Hemelwater, het riool in?* [Rainwater, into the Sewer?], Faculty of Civil Engineering, Technical University of Delft, Delft (NL).

National Institute of Freshwater Management and Waste Water Treatment (2002) *Rol van provincies in stedelijk waterbeheer: Resultaten van indicatief onderzoek* [The Role of Provinces in Urban Water Management: Result of Indicative Research], National Institute of Freshwater Management and Waste Water Treatment, Lelystad (NL).

Province of Drenthe (1998) *Provinciaal Omgevingsplan Drenthe* [Provincial Comprehensive Plan], Province of Drenthe, Assen (NL).

Province of Drenthe (1999) *Besturen in balans: Bestuursprogramma 1999-2003* [Balanced Administration: Administrative Programme 1999-2003], Province of Drenthe, Assen (NL).

Rathenau Institute (2000) *Het blauwe goud verzilveren: Integraal waterbeheer en het belang van omdenken* [Silverplating the Blue Gold: Integrated Water Management and the Importance of Reconsidering Water Management], Rathenau Institute, The Hague.

Toonen, T.A.J. (2000) *Een bestuursschouw van het waterschap* [An Administrative Review of Dutch Water Boards], Interprovincial Council, The Hague.

V&W (Ministry of Transport, Public Works and Water Management) (1985) *Omgaan met water: naar een integraal waterbeleid* [Dealing with Water: Towards an Integrated Water Management], The Hague.

V&W (Ministry of Transport, Public Works and Water Management) (1990) *Derde Nota Waterhuishouding* [Third Memorandum on Water Management], Ministry of Transport and Public Works, The Hague.

V&W (Ministry of Transport, Public Works and Water Management) (1998) *Vierde Nota Waterhuishouding* [Fourth Memorandum on Water Management], Ministry of Transport, Public Works and Water Management, The Hague.

Chapter 14

An Evaluation of Sustainable Housing Policy in a Trans-Provincial Region of the Netherlands

Geoff Porter and Frans Osté[1]

14.1 The role of governmental actors and stakeholders in representing community needs

Stakeholders use many different means to achieve their objectives and to influence policy outcomes (Klijn, Koppejan, and Termeer 1993). However, it is usually necessary for them at some stage of the decision-making process to have to justify their claims by reference to the interests of the wider community (Lafferty and Meadowcroft 1996). For example, a planning proposal for a housing development scheme might be justified by the need to attract high-income earners into the area, despite a potentially significant environmental impact. This conflict of interests between economic development and its social and environmental consequences lies at the heart of the case study in this chapter. It examines the wider implications for spatial policy for housing, by reference to the present, desired and potential contributions of a wide range of actors.

As already discussed throughout this book, current debates on 'sustainability' reveal that individuals and groups might hold completely different perceptions of the nature of this seemingly straightforward notion. Decision-making with regard to sustainability is not always fully transparent however, and this can sometimes stand in the way of the community interest. The question of what serves the community interest can take on a variety of differing interpretations. Should it serve the local interest economically, environmentally, or are other aspects more desirable? And the local interest might, for example, conflict with the regional interest (see Chapter 4). Thirdly, we have the question of who should represent the interests of the community. Local government might in some cases, for example, act both as an advocate of the public interest and a defendant of its own interests (Kaiser 1995). Municipal interests in the real estate market are a case in point. A problem is that governments

---

1 The late Geoff Porter was a researcher at Sustainable Cities Research Institute, at the University of Northumbria, Newcastle upon Tyne; Frans Osté is a researcher at the Department of Planning and Environment, Faculty of Spatial Sciences at the University of Groningen, the Netherlands.

can become to some extent detached from society, and as such they might fail on occasions to represent the aspirations of citizens (Healey 1997).

In order to support a transition towards more sustainable policy and practice, this book has argued extensively that the perceptions and motives of a wide range of actors must be clearly understood. Those actors who stand to lose out in conflicts between for example environmental protection and economic development will take action to defend their interests. This will hamper contributions to sustainable development, whatever its definition. Furthermore, such processes may prevent politicians from lending their support to an environmental cause, because such action could compromise their political survival. Consequently, it can be difficult to convert genuine intentions in the field of sustainable development into concrete action.

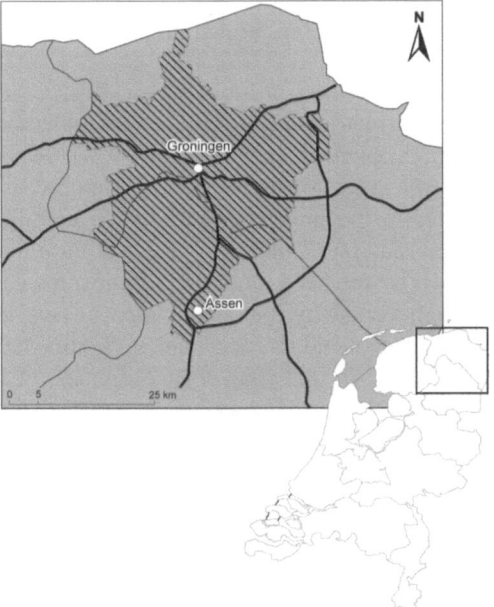

Figure 14.1   The Region-Vision area

It can be concluded that two aspects are important when addressing 'sustainable development' in planning policy. The first point concerns the need to understand the underlying mechanisms that hamper actor contributions to sustainable development. A better understanding of these issues might assist in the improvement of the institutional setting and in policy formulation. The second point refers to the need for wide-ranging debate about the meaning of particular sustainable development issues. Hajer (1995) argues, for example, that environmental issues should not be conceptualised in terms of defined, unambiguous concepts by individual actors. Rather this conceptualising or reframing of reference will entail a continuous struggle and exchange of ideas over the definition and meaning of the issue. This chapter describes how these considerations were incorporated into a case study of housing policy, in the region of Groningen-Assen in the north of the Netherlands (see also Buiten et al. 2001).

14.2 Sustainable housing in the Groningen-Assen Region

Covenants are important instruments for establishing regional strategic plans and implementation agreements. An example is the 'Vision for the Groningen-Assen Region', which is a non-statutory regional plan including an agreement between the Provinces respectively of Groningen in the north and Drenthe in the south, and twelve municipalities (see Figure 14.1). The functional coherence of this area is influenced mainly by the relationship between the two cities of Groningen and Assen and their surrounding area. The population in the city of Groningen is approximately 171,000, while the city of Assen has around 56,000 inhabitants. The entire Groningen-Assen area accommodates more than half a million people.

Several reasons underly the need for co-operation among the actors in this functional region. For instance, many people are employed in the two main cities, but live in the smaller settlements around the towns. There is also a substantial interaction between the two cities, as the inhabitants continuously use each other's facilities. Under these circumstances, a municipal approach will obviously fail to tackle the consequences of these behavioural features.

Co-operation and co-ordination have also become necessary in the face of stiff competition among municipalities in their effort to attract businesses, generate economic activities, and attract middle and higher income households into their respective municipalities. This competition has resulted in out-migration of higher income groups from both cities. This has caused problems especially in Groningen city's post-war neighbourhoods. This segregation process has also adversely affected the fiscal income of especially the city of Groningen. Apart from the social and economic drawbacks, the out-migration of middle and higher income households from the cities has resulted in ongoing growth and development of surrounding villages and towns. This process of suburbanisation has in turn caused a deterioration of vulnerable types of land use, including for example nature areas and water systems.

In other words, both environmental and economic problems have occurred due to lack of regional policy co-ordination. The local interest of municipalities often prevails above the collective interests of the region. In this case housing market distortions and the transformation of green areas into built-up areas have been the result. These difficulties illustrate the background to this review of regional policy for sustainable housing within the Groningen-Assen region.

14.3   The research methodology

Confronted with the environmental, economical and – as a result – social complications of a distorted housing market, government authorities in the Groningen-Assen region desired a better insight into an uncertain policy arena. All had in common the concern that communities within the region were becoming unbalanced. It was considered that the opinions of a small number of actors, who were easy to identify, were crucial in gaining a better understanding of this policy arena. Arguably, these are the conditions that beg for the 'actor-consulting' model. The key aim of the model is to develop an understanding of the motivation of influential actors in the relevant field of policy, and to unravel underlying mechanisms that determine their actions.

It should be noted at this point that the research was carried out on behalf of the Province of Drenthe. The consequent opportunities for developing regulation to improve sustainability of housing developments would therefore in the first instance be confined to action taken by Drenthe, although clearly the results of the study would be of interest to all of the provincial and municipal organisations in the region.

The research methodology started with activities to establish the *present* and *desired* contributions of each of the actors. The second phase focused on the development of *potential* contributions of actors. The third phase sought to gain an insight into the way in which the actors might respond to the initial ideas for potential contributions, as a means to develop a a more sustainable way forward for regional housing policy.

The following process was employed to investigate the *present* and *desired* contribution of actors involved. Firstly, a contextual study was carried out to establish the current regulatory framework for housing location and design, in terms of direct, indirect, and self-regulation, along with the historical context underlying the formulation of current policy. Secondly, an investigation of the authority's own perception of its policy environment was carried out to establish their expectations for the roles of the different parties, and their perception of the balance of power among the various actors. Thirdly, a questionnaire was devised to explore each of the actors' present and desired contributions to housing location and design, and interviews were carried out to explore any discrepancy between the desired and present contributions.

After analysing the present and desired contributions, research was carried out to ascertain the *potential* contribution of the actors. This was achieved in the first instance by carrying out a comparative study between the Groningen-Assen region and a similar situation in other regions. This enabled the researchers to establish *what solutions* might exist. The comparative study was then supplemented by interviews with actors to ascertain their views on how these potential contributions might help to achieve mutually beneficial objectives. This provided a better understanding of the *advantages and disadvantages* of these solutions for each individual actor. The outcome was an outline set of policy instruments, or potential contributions, which could be used as a basis for a revised regulatory regime.

The final activity sought to obtain an insight into the way in which parties might respond to the proposed policy ideas. In an ideal world, the best outcome of the process so far would be complete agreement among parties about the effectiveness of the proposed policy. In reality though, the outline policy ideas require adjustment to obtain an optimum balance. Consultation work was therefore carried out to establish how each actor might behave in response to the proposed ideas for regulation. This information was used to adjust the proposed regulatory regime, to recognise more fully the relationships between the decision-makers and the executors of the decisions.

14.4 Analysis of actor contributions

The aim of the analysis work was to explore a better understanding of the underlying motives of actors in the Region Vision area. The most important actors involved in regional housing and land use planning are developers, social housing corporations, consultants, and a variety of interest groups. The national objectives for housing are aimed at concentrating the population in the larger cities, while discouraging housing development in rural areas. This is the policy frame in which all regional parties are supposed to work. As such, the policy of encouraging housing redevelopment on derelict brownfield land supports the containment of urban sprawl and out-migration from the cities. The more specific policy in Drenthe for the reconstruction of post-war neighbourhoods (see Figure 14.2) is intended to reverse the trends of decreasing occupancy in urban areas, where the quality of the housing is failing to match expectations.

In the past, the Province of Drenthe pursued these objectives mainly through deploying planning instruments derived from formal legal procedures. These instruments include the zoning of land use, and controlling the number of houses that each municipality is allowed to build.[2] These instruments, which enable the Province to influence the supply of houses, mainly address such questions as 'how many' dwellings ought to be built, and 'where'. Less attention has been paid to the actual *demand* for housing, and to the questions of 'what is built' and 'what needs to be built'. So far, the provincial policy has been relatively top-down and passive in its nature and mainly oriented towards guiding and restricting the municipalities. This emphasis has left the responsibility for meeting the individual housing demand, and for the quality of the developments, to local actors.

---

2   Dutch spatial planning law obliges provincial authorities to draw up a regional land use plan for the area under their jurisdiction. This plan, among others, forms the basis for evaluating local plans. The regional plan includes a rough sketch of the zoning of new housing developments. The Groningen-Assen region is included in two regional plans: the southern part is in the regional plan of Drenthe, whereas the northern part falls under the regional plan of the Province of Groningen. However, both plans take into account the mutual agreements in the 'Vision of the Groningen-Assen Region'.

Figure 14.2   Reconstruction site of a post-war neighbourhood

The interests of these actors, and their consequent actions, do not always contribute to more sustainable housing in the region. For example, it is common practice in the Netherlands for *municipalities* to acquire greenfield land, provide the infrastructure, and subsequently sell the land at a profit to developers, housing corporations or individual households. Part, if not all of the profit is used to provide the area with an infrastructure. The new landowner will then build houses according to the regulations of the local land use plan. Financial interests, linked to the ownership of the land, are a driving force behind the location of new developments, and these financial considerations often take priority over considerations of environmental and social sustainability.

The actor consulting model revealed a failure to recognise actors' desires and the needs of the community as a whole. Whereas rural municipalities in the Region Vision area have made considerable financial profits from converting agricultural land, the urban municipalities have found that the reconstruction of their post-war neighbourhoods and the redevelopment of brownfield land have resulted in substantial financial losses. Furthermore, private developers have not queued up to invest in the latter type of project, since reconstruction projects go hand in hand with uncertainty.[3] This situation has created a stimulus to work towards more common goals, including mechanisms supporting balanced regional housing distribution.

As stated before, the provincial policy could and will partly influence the building activities of municipalities. However, the traditional regional plan of Drenthe is rather limited in its possibilities, as it does not prescribe the 'target groups' for housing developments, or qualitative aspects of new housing such as the size of houses. As such, a block of flats for the elderly – a group with a high demand for housing – may consume a large part of a municipality's housing allocation. Therefore, satisfying the housing demand of the elderly does not rank high on the agenda of rural municipalities. Instead, it is more financially attractive to use their housing allocation to build traditional family dwellings for middle and higher income groups. More people can live in larger family houses than in smaller apartment blocks, of which one unit is counted under the terms of the regulations as the equivalent to one family dwelling.

The traditional provincial planning system did not promote a change in the type of land use. A change from derelict brownfield land to residential neighbourhood might not be viable, because the costs outweigh the potential returns. New high-density housing is often necessary on these sites to make such projects financially feasible. However, high-density housing is often not possible due to adverse market demand, and considerations of design quality. High densities have in many instances failed to support the quality or the sense of identity of places, particularly in rural municipalities. The actors tended to associate high-density designs with lack of public space, poor layout design, disjunction with the existing townscape, and lack of attention to ecological measures. This provides a motive to work towards a collectively supported financial program in support of re-use of brownfields in the region.

In the past municipalities have developed sites themselves and some were inclined to relax building restrictions in order to sell parcels of land. Nowadays,

---

3   Uncertainty with regard to the development of brownfield land may relate to such issues as soil contamination, the reaction to development proposals of local residents, or the difficulty in predicting land acquisition costs or the final value of the real estate.

it is gradually becoming common practice for municipalities to withdraw from the development process by transferring an entire new housing allocation area to one or several project developers, while ensuring that certain quality standards are adhered to. In the former approach the municipality earned full profit from the development projects. In the market-oriented approach the developers stand to profit the most. Unfortunately, it remains a problem that instruments for prescribing quality, such as policies for securing the release of public space, are not sufficiently developed or properly applied by municipalities. A region-wide policy framework that sets out commonly agreed quality regulations is needed to deter developers from their current practice of shopping around for the best greenfield sites, and design proposals that yield the highest profits.

14.5   Recommendations for regulation to improve sustainable development[4]

In order to begin the process of developing recommendations for improved regulation of housing development, it was firstly necessary to examine the existing objectives of the Province of Drenthe for the distribution of housing. These cover many issues such as the need for specified quantities of housing, quality requirements, intrusion in the landscape, and sense of identity. This can be considered as the *present contribution* of the Province, whereas their *desired contribution* is to establish a more coherent set of policies to meet these objectives.

The *organisation* – the Province – mainly applies traditional instruments that are derived from formal legal planning procedures. Most of these instruments are *direct regulations*, and are mainly focused on 'how many' houses may be built and 'where'. In other words, the provincial policy focuses on its traditional role – the protection of vulnerable areas against highly dynamic requirements for land use.

Meanwhile, other actors' *present contributions* do not always support sustainable development of housing in the region. Their actions are mostly motivated by financial interests, which are not always in line with the regional interest. Reconstruction of post-war neighbourhoods in the larger towns involves a substantial financial commitment, whereas the undesired expansion of housing on agricultural land brings in money. Smaller municipalities are limited in their building activities and will not build for the elderly as much as they should. Also project developers prefer investing in new residential areas, but do not gain from extra investments in quality, because after the completion of the project, the owners who sell their property appear to benefit most.

Money determines to a large extent the motivation and power for influencing *desired developments* that contribute to sustainable housing (See also Forrester 1989). Clearly, there is much to gain if common policy can be composed. For the Province of Drenthe this means a shift in approach and attitude towards the

---

4   The italic print in this section refers to the model described in Chapter 8.

various regional (municipal) actors, from direct and coordinative regulation towards supporting common interests on the basis of consensus and mutual understanding.

There are opportunities – *potential contributions* – that might support the sustainability objectives. The allocation of housing numbers is not sufficient. Instead of controlling 'how many' houses are built and 'where', the Province should also pay attention to 'for whom' and 'what' to build. This can be controlled by *direct regulation*, by a clearer prescription of the type of housing required, avoiding for example, the current practice of accounting for the supply of a single-bedroom apartment in the same way as a family dwelling.

Some benefit might also be derived from guidelines for quality, identity and ecological building, especially in instances of public-private co-operation. Nevertheless, full control through direct regulation is not desirable since that would exclude flexibility in municipal actions. Instead, the regional authorities should apply more *indirect regulation* in order to influence local actors when their decisions affect the regional interest. In such instances the regional government has to pursue an active policy that focuses on 'development' instead of passively supporting 'protection' in land use planning. However, an increase in regional budgets that would be needed to support this approach is unlikely, since a transfer of financial responsibility from the local to the regional level is not realistic.

Consequently, in the absence of any prospect of regional fiscal support, the Province might achieve more sustainable housing outcomes by articulating the regional interest, trying to involve the parties that affect the regional interest, and looking for solutions that meet the needs of the population in a regional context. Crucially, the Region Vision covenant between the two Provinces and twelve municipalities provides a promising mechanism to achieve this aim, as it has the support of a wide variety of partners.

The case illustrated in this chapter has implications for the planning of housing not only in the Groningen-Assen region. Traditionally, throughout western Europe, spatial planning has tended to focus on housing *supply*. Responsibility for housing *demand*, and in particular the provision of affordable housing, has usually been separated from the spatial planning department across local, regional and national governments. By using actor consulting to explore a wider spectrum of issues among both government and private sectors, a more holistic approach can be envisaged to developing governance processes that lead to a more effective balance between housing supply and demand.

14.6 Commentary on the methodology

As outlined in the opening section, it had been anticipated that two aspects would prove to be important when developing ideas for regulation to improve the sustainable development of housing in the region. The first point concerns the importance of determining the underlying mechanisms that hinder actions that support sustainable housing development. The second point refers to the need for a wide-ranging debate about the meaning of sustainable development, which in this case relates to how housing can be developed on a sustainable basis throughout the region.

The results of the research appear to be in line with the first point since an insight was readily obtained of the actors' motivations with regard to sustainable development, and the problems they face when addressing sustainability issues. Barriers to achieving more sustainable practice were readily identified, without recourse to detailed discussions on the meaning of 'sustainability'. With regard to the second point, the methodology has been successful in uncovering thoughts, opinions, and expectations about sustainable development. The approach, however, does not replace participation processes in which actors can exchange ideas and thoughts: rather, it enables the controlling authority to obtain a better understanding of the opportunities and difficulties in developing improved regulation. The results obtained may therefore act as an ongoing means to develop a more structured and informed planning dialogue by the Province of Drenthe and its regional partners.

It has been of significant benefit to explore policy options in the context of the functional region of Groningen-Assen, rather than within the confines of any Provincial border. This has helped to avoid 'social dilemma' problems on a regional scale, and to develop policy ideas that are helpful to both of the Provincial governments. The position of the municipalities has been clarified, as they now have the opportunity to adopt a mutually more consistent approach to housing policy.

Actor-consulting appears to have been of particular value in a situation typified by uncertainty amongst actors with regard to the objectives for regional housing development. The method has facilitated a decrease in uncertainty, by identifying shared meanings, by identifying barriers to more sustainable development, and by gaining consensus on opportunities for action. Sustainability is not given a better definition, but the case illustrates the need for more thoughtful decision-making and a better understanding of how actors and their desires need collective actions: Sustainability is in this case synonymous with interaction towards a mutual understanding and balancing objectives. In many ways, the *process itself* has been an important aspect of the method, in facilitating improved communication.

References

Buiten, J., W. Huizing, F. Osté, G. de Roo and H. Voogd (2001) *Regional Planning and Sustainable Development: Towards an actor-oriented approach: Final Report Dutch Case Study SUSPLAN project*, University of Groningen / Province of Drenthe, Groningen / Assen.

Forester, J. (1989) *Planning in the face of power*, University of California Press, Berkeley, USA.

Hajer, M. (1995) *The Politics of Environmental Discourse: Ecological Modernisation and the Policy Process*, Oxford University Press, Oxford.

Healey, P. (1997) *Collaborative Planning: Shaping Places in Fragmented Societies*, Macmillan Press, London.

Kaiser, E.J., Godschalk, D.R. and F.S. Chapin (1995) *Urban Land Use Planning*, University of Illinois Press, Chicago.

Klijn, E.H., Koppejan, J.F.M. and C.J.A.M. Termeer (1993) 'Van beleidsnetwerken naar netwerkmanagement' [From policy networks towards network management], in: Beleid en Maatschappij, Vol. 5, pp. 230–242.

Lafferty, W.M. (1996) 'The Politics of Sustainable Development: Global Norms for National Implementation', *Environmental Politics*, Vol. 5, No.2, pp.185–208.

Lafferty, W.M. and J. Meadowcroft (1996) 'Democracy and the environment: congruence and conflict – preliminary reflections', in: W.M. Lafferty and J. Meadowcroft (eds), *Democracy and the Environment: Problems and Prospects*, Edward Elgar, Cheltenham, UK.

Index